Nanotechnologies, ethics and politics

The ideas and opinions expressed in this publication are those of the authors and are not necessarily those of UNESCO and do not commit the Organization.

The designations employed and the presentation of material throughout this publication do not imply the expression of any opinion whatsoever on the part of UNESCO concerning the legal status of any country, territory, city or area or of its authorities or concerning the delimitation of its frontiers or boundaries.

Published by the United Nations Educational, Scientific and Cultural Organization
7, place de Fontenoy, 75352 Paris 07 SP, France

Typeset by UNESCO Publishing/Roberto Rossi
Printed by Darantiere, Quétigny

ISBN 978-92-3-104051-1

Printed in France

NANOTECHNOLOGIES, ETHICS AND POLITICS

Edited by Henk A. M. J. ten Have

Ethics Series

UNESCO Publishing

Contents

Preface

The UNESCO Member States established the World Commission on the Ethics of Scientific Knowledge and Technology (COMEST) in 1998. Composed of eighteen leading experts from various countries and with different professional expertise, COMEST advises UNESCO on the ethics of science and technology (S&T). One area of focus is on environmental ethics, which has been treated in a previous volume in UNESCO's Ethics Series (*Environmental Ethics and International Policy*, 2006). Another area of focus is on science ethics; here COMEST examines the fundamental principles of science and research, and analyses and provides information on codes of conduct for scientists in various parts of the world and in various professions and disciplines. A third area is the ethics of technology. COMEST is working, for instance, with space agencies to draw attention to ethical issues raised by space technology and space exploration. A new subject in this third area concerns nanotechnology, namely the ethical dimensions of nanotechnologies. It is with this subject that the present volume is concerned.

As an international ethics body, COMEST's mandate is (1) to be an intellectual forum for the exchange of ideas and experience, (2) to thereby detect early signs of risk situations, (3) to advise decision-makers in view of such situations, and (4) to promote dialogue between scientific communities, decision-makers and the general public. In pursuing its mandate, COMEST has analysed information technology, water use and hydrological technologies, energy and space technology.

NANOTECHNOLOGIES, ETHICS AND POLITICS – ISBN 978-92-3-104051-1 – © UNESCO 2007

Nanotechnology is one of the most rapidly developing technologies today. All experts agree that – like nuclear technology, genetic technologies, biotechnology and information technologies before it – nanotechnology will sooner or later have an enormous impact on societies and human life. Its impact will be even stronger than that of its predecessors since nanotechnology combines previous technologies and bears on many areas of human life and action. Since we are on the verge of a new technological revolution, now is the most appropriate time to assess the ethical implications of the new technology. While recognizing that the assessment can only be preliminary, since it is impossible to predict with precision how S&T will evolve, and that ethics is often felt to offer too little too late concerning developments in S&T, we nevertheless find ourselves presented with the unique opportunity to engage in prospective (rather than the usual retrospective) ethical reflection on nanotechnology. Such reflection aims to provide guidance as the new science develops, thereby supporting arguments that developments should accord with certain significant human values and be aimed at redressing fundamental global problems.

It is a mistake to assume that ethical reflection on nanotechnology is not relevant on a global scale. It is striking that several developing countries are investing heavily in nanotechnology. Countries such as China, India, Brazil, South Africa and Mexico have extensive research programmes in this area. The risk is that with the rapid growth of nanotechnologies, the divide between more- and less-developed countries will not only deepen but also be transformed in character. Like every new technology, nanotechnology should yield benefits for all human beings wherever they live. All countries should be involved in such developments so as to partake of the benefits of S&T. Another risk, however, is that developing countries will function only as testing sites and provide subjects for experiments, thereby reaping not the rewards of such research but only the ills, disadvantages and adverse effects thereof. An important issue regarding new advances in S&T is thus the basic ethical question of how they can be directed towards the amelioration and solution of global problems, such as poverty, environmental degradation, violence and disease. Precisely for this reason, reflection on ethical issues at this preliminary stage is crucial for such a body as UNESCO.

The present volume provides, first of all, information on nanotechnology: What is the state of the art in S&T? It also elaborates

NANOTECHNOLOGIES, ETHICS AND POLITICS – ISBN 978-92-3-104051-1 – © UNESCO 2007

and analyses the ethical issues related to nanotechnologies. Finally, it addresses policy issues. COMEST is developing policy advice on the ethics of nanotechnology with recommendations for international action by UNESCO. This book is therefore a starting point for further exploration of ethical issues. It is also intended to stimulate interdisciplinary dialogue not only between scientists and ethicists but also with policy-makers, the general public and special interest groups.

I thank the contributors for their efforts to clarify the ethical issues involved in nanotechnologies. It is my hope that this book will significantly increase awareness of the issues and stimulate further debate about the ethics of nanotechnologies.

Pilar Armanet
Chairperson
World Commission on the Ethics of Scientific Knowledge and Technology (COMEST)

Contributors

Kyunghee Choi

Ehwa Woman's University, Korea Daehyon-Dong, Seoul, Republic of Korea

Erin Court

McLaughlin-Rotman Centre for Global Health, Program in the Life Sciences, Ethics, and Policy, University Health Network and University of Toronto. Currently at the University of Oxford, UK

Abdallah Daar

McLaughlin-Rotman Centre for Global Health, Program in the Life Sciences, Ethics, and Policy, University Health Network and University of Toronto

Edith Deleury

Chair of the Quebec Commission on Ethics in Science and Technology (CEST), Quebec, Canada

Diane Duquet

Coordinator, Quebec Commission on Ethics in Science and Technology (CEST), Quebec, Canada

Bert Gordijn

Department of Ethics, Philosophy and History of Medicine, Radboud University Medical Center, Nijmegen, the Netherlands

Donald Evans

Bioethics Centre, Dunedin School of Medicine, University of Otago, New Zealand

Henk ten Have
Division of Ethics of Science and Technology, UNESCO, Paris, France

Michele Jean
Faculté des études superieurs, University of Montreal, Montreal, Canada

Jixing Liu
Institute of Theoretical Physics, Chinese Academy of Sciences, Beijing, People's Republic of China

Fabio Salamanca-Buentello
McLaughlin-Rotman Centre for Global Health, Program in the Life Sciences, Ethics, and Policy, University Health Network and University of Toronto

Joachim Schummer
Department of Philosophy, Technical University of Darmstadt, Darmstadt, Germany

Peter A. Singer
McLaughlin-Rotman Centre for Global Health, Program in the Life Sciences, Ethics, and Policy, University Health Network and University of Toronto

Margareth Spangler Andrade
Fundaçao Centro Tecnologica de Minas Gerais, Belo Horizonte, Brazil

Introduction

UNESCO, ETHICS AND EMERGING TECHNOLOGIES

Henk ten Have

On 16 November 1945, representatives from thirty-seven countries meeting in London decided to establish an organization that would create a genuine culture of peace through education, science, culture and communication. The constitution of this new organization was to give hope to the international community for a world of solidarity and dignity after the war that had just ended. The United Nations Educational, Scientific and Cultural Organization (UNESCO) is the only UN agency with specific responsibilities for scientific research. This implies promoting scientific collaboration among the peoples of the world in order to advance the objectives of international peace and the common welfare of humankind. It also implies making sure that the advances of science and technology (S&T) will take place within the framework of 'universal respect for justice, for the rule of law and for the human rights and fundamental freedoms which are affirmed for the peoples of the world', as it is expressed in Article 1 of UNESCO's Constitution (1945). The ethical aspects of scientific research therefore need to be considered. In the 1980s, UNESCO focused activities on the human genome project and its ethical implications. In 1989, the General Conference invited the Director-General of UNESCO to introduce a permanent system of consultation for the exchange of information and experience on the ethical implications of contemporary S&T (UNESCO, 1989). The aim was to make UNESCO a clearing house for information and documentation on the ethics of S&T within the United Nations system in order to promote 'forward-looking reflection'.

Since then, UNESCO has been involved in organizing and sponsoring international activities in bioethics (see, e.g., Sass, 1991). In 1993, the Member States requested that the Organization consider the feasibility of establishing an international legal framework for the protection of the human genome. The motivation for this request was not only the desire to assure that due regard be given to the freedom, dignity and identity of the human person, but also the need to ensure the participation of all in the advances of the biomedical and life sciences and in the resultant benefits (UNESCO, 1993). In the same session, the Member States decided to establish the UNESCO International Bioethics Committee (IBC). This has been the start of an explicit programme in the ethics of S&T, bioethics in particular. In 2002, ethics was earmarked as one of the five chief priorities of UNESCO.

The current revolution in S&T has led to the concern that unbridled scientific progress is not always ethically acceptable, and at least requires careful ethical reflection. The need to establish common values and benchmarks for all countries alike, as well as to promote ethical principles and standards to guide scientific progress and technological development, is becoming increasingly acute, especially in developing countries that do not equally enjoy the benefits of advances in S&T. UNESCO's work in the ethics of S&T reflects these global concerns as rooted in the cultural, legal, philosophical and religious heritage of the various human communities. This is another characteristic of present-day ethics: it is not only related to science and scientific developments but also has an essentially cultural and educational dimension. Finally, the priority of ethics in UNESCO reflects the fact that ethics of S&T is increasingly considered a responsibility of the world's political community (Lenoir, 1997).

LINKING SCIENCE AND POLICY: COMEST

The UNESCO ethics programme was expanded beyond the domain of bioethics in 1998 with the establishment of the World Commission on the Ethics of Scientific Knowledge and Technology (known under its French acronym, COMEST). This commission is composed of eighteen prominent and independent scientists and other experts from different regions of the world and from various scientific disciplines, including education, engineering, history, law, mathematics, philosophy, politics and the social sciences. It is responsible for advising the Organization on issues concerning the ethics of S&T. The Secretariat of COMEST and the International Bioethics Committee (IBC) is located in UNESCO's Division of Ethics of Science and Technology in the Sector for Social and Human Sciences. COMEST has the mandate to be an international advisory body and an intellectual forum for the exchange of ideas and experience, as well as to encourage the scientific community to examine fundamental ethical issues – and to detect early signs of risk situations. It formulates ethical principles that can shed light on the various choices and impacts brought about by new discoveries. It advises decision-makers on policy issues and promotes dialogue between the international scientific community, governments and the general public concerning sensitive areas (such as sustainable development; freshwater use and management; energy production, distribution and use; outer space exploration and technology), as well as issues of rights, regulations and equity related to the rapid growth of the information society. COMEST executes its mandate by bringing together experts who study specific problems and disseminating the results of their analyses through publications. Ethics and space technology, ethics and energy, and ethical issues in relation to water use are areas that have been examined in the past and have led to widely disseminated publications (e.g. Pompidou, 2000; COMEST, 2004).

One recent publication concerns the precautionary principle. Because this principle is controversial from an international perspective, a group of experts analysed the concept and its applications in diverse settings in order to clarify possible misunderstandings. The resultant publication (COMEST, 2005a) is the first concerning environmental ethics in general. COMEST is preparing policy advice to identify possible areas of work for UNESCO. Many UN agencies are active regarding environmental issues. However, environmental ethics is not typically explicitly addressed. To place ethics on the international agenda, it will first be necessary to develop a careful approach, cognisant of the fact that environmental ethics, compared to bioethics, is a more recent area of applied ethics and that many controversies and disagreements exist concerning its basic principles.

The same considerations apply to the ethics of technology, the most recent aspect of which concerns nanotechnologies. Although COMEST has been working on ethics in relation to space technologies, the presentation here is the first to address the domain of new and emerging technologies.

To continuously update its global perspective, COMEST organizes a public session every two years, bringing together for several days scientists, ethicists, lawyers and policy-makers to discuss salient ethical issues in S&T. Such well-attended conferences are organized in different regions of the world both to provide a platform for global concerns and to stimulate the ethical debate and the creation of networks of experts in those regions. Recent conferences have taken place in Rio de Janeiro (COMEST, 2003), Bangkok (COMEST, 2005b) and in Senegal (December 2006). To coordinate the increasing number of international activities, especially in bioethics, the Director-General of UNESCO established the United Nations Inter-Agency Committee on Bioethics in 2003. The committee involves intergovernmental organizations both within and outside the UN system, such as the World Health Organization (WHO), the Food and Agriculture Organization (FAO), the International Labour Organization (ILO), the United Nations High Commissioner for Human Rights (UNHCR) and the World Intellectual Property Organization (WIPO), but also the Council of Europe, the European Commission (EC), the Organisation for Economic Co-operation and Development (OECD), and the Arab League Educational, Cultural and Scientific Organization (ALECSO). The committee provides a forum for debate and exchange of information, promotes coordination of activities, facilitates collaboration and engages in concerted action (such as examining the ethical issues regarding intellectual property rights). The committee meets at least once each year. It reports to the Secretary General of the United Nations and to the executive heads of the participating organizations.

UNESCO's activities in the ethics of S&T therefore take many forms and cover much ground. They include, for example, drawing up recommendations for decision-makers and drafting and developing ethical guidelines, standards and legal instruments. UNESCO also helps to develop regional networks, to build and expand national capacities, to promote ethics in science education, and to provide educational materials. Further, it performs an essential 'ethical watch' function and plays an important role as a catalyst and think tank, informing public opinion on the human rights implications of progress in S&T.

The activities thus far have fallen within three categories: (1) standard-setting: developing internationally agreed-upon normative frameworks to guide policy-making in Member States; (2) capacity-building: enhancing the ethics infrastructure in Member States so that they will be better able to cope with ethical issues emerging from the application and development of S&T; and (3) awareness-raising: supporting and promoting public debate on ethical issues so that as many stakeholders as possible will have a better understanding of the moral choices at stake in globalized world.

STANDARD-SETTING

As a consequence of the global dissemination of S&T, ethics has become increasingly international in scope. Ever more research takes place in multiple centres and internationally, with more and more research subjects being recruited in developing countries. Healthcare practices have become global while guidelines and legal contexts differ and at times are even absent. Rules for transplantation and procedures for organ donation vary among countries and these different approaches have led to abuses, such as organ trafficking and commodification of transplantation practices. Moreover, the burdens and benefits of advances in S&T are not equally distributed. Poorer countries risk being excluded from the benefits of biomedical progress. There is also a risk that double, or at least different, moral standards will be applied in different regions of the world. While developed countries are increasingly establishing regulations and passing legislation concerning scientific development, technological innovation and environmental protection, less developed countries lacking such legal frameworks run the risk of becoming 'dumping grounds' for toxic waste or 'testing grounds' for risky experiments. The recent scandal over toxic waste in the Ivory Coast and the movie *The Constant Gardener* provide illustrations of the dangers of inequalities in legal and regulatory frameworks.

Many countries have only limited bioethics infrastructures, for they lack expertise, educational programmes, bioethics committees and legal frameworks. The global nature of S&T implies the need for a global approach to ethics. Member States have mandated UNESCO to set universal ethical benchmarks covering issues raised by the rapid development of S&T. Those states want to work together to identify basic principles and shared values regarding S&T. Standard-setting action in ethics has become a necessity that is felt throughout the world, often expressed by scientists and practitioners themselves, as well as by legislators, policy-makers and citizens. At the same time, the involvement of an intergovernmental organization in ethics is controversial. It is argued that governmental regulation, and thus political interference, is unwanted since ethics should be left to civil society, professional organisations and public debate. Organizations such as UNESCO should not regulate, dictate or impose any ethical principles. On the other hand, nowadays many governments have formulated ethical regulations; quite a few countries, especially in Europe, North America and Latin America, have drafted 'bioethical legislation', for example – particularly regarding medical research and reproductive medicine. Such governments may be anxious that ethical agreements at the level of UNESCO might contradict or be inconsistent with their domestic legislation. Nonetheless, it is pointed out in every intergovernmental debate that UNESCO, as an organization with 192 Member States, has indeed a role to play since the majority of its members do not have an adequate ethics infrastructure, and often lack any regulation and legislation. The desire to develop international frameworks is therefore often articulated by the least developed countries that are in need of normative guidance and that want to have the certainty that ethical principles are formulated on a global level so that the same standards are used everywhere.

It was in this context that in October 2003, based on preliminary feasibility studies of IBC, UNESCO was mandated by its Member States to draw up a declaration setting out fundamental principles in the field of bioethics. After two years of intense work, these same Member States adopted, unanimously and by acclamation, on 19 October 2005, the *Universal Declaration on Bioethics and Human Rights*, thus solemnly affirming the international community's commitment to respect a certain number of universal principles with regard to humanity in the development and application of biomedical S&T (UNESCO, 2006c). Previously, the Member States had adopted normative instruments in the more specialized area of genetics, namely the *Universal Declaration on the Human Genome and Human Rights* (1997) and the *International Declaration on Human Genetic Data* (2003).

CAPACITY-BUILDING

The objective of the UNESCO programme is to identify ethical issues that are relevant to the various regions of the world so as to determine and implement appropriate strategies both for encouraging ethical reflection at regional and sub-regional levels and for strengthening national capacities and international cooperation in ethics. If these efforts are to be successful, it is essential that the legal, cultural and religious traditions of each Member State be taken into account. However, standard-setting activity is only the first step towards achieving the programme's objective. The adopted declarations will remain mere paperwork so long as their provisions are not taken into account by and put to work in each Member State.

However, the ethics infrastructure in Member States is extremely heterogeneous – varying from countries with many ethics experts, teaching programmes, legislation and ethics committees, to countries with hardly any of these. This heterogeneity also means that the input in the ethical debate often comes from wealthy developed countries alone. In the case of the scientific bioethical literature, for instance, researchers from developing countries contribute little (Borry et al., 2005). UNESCO is undertaking major efforts to assist Member States in building and reinforcing their ethical infrastructures. In particular, its three main activities are to (a) provide factual information, (b) promote ethics teaching and (c) establish (bio)ethics committees.

Global Ethics Observatory (GEObs)

To provide Member States with proper tools for reflection and appropriate means for coping with emerging ethical challenges in S&T, the Global Ethics Observatory (GEObs) was launched in December 2005 (www.unesco.org/shs/ethics/geobs). GEObs currently consists of three databases. The first database ('Who's who in ethics?') presents data on experts in various areas of ethics. A questionnaire has been developed and mailed to experts in all regions of the world. The database allows one to search for different types of experts by country, area of expertise, experience and keywords. The second database includes data on institutions such as ethics committees (at different levels: local, national, regional, international), ethics departments and centres, and ethics associations and societies. This database, as the others, will cover all areas of applied ethics, including bioethics, nursing ethics, law and ethics, social sciences and ethics, science ethics, environmental ethics and engineering ethics. It will also present, in due course, all data in the six official languages of the Organization. The third database presents descriptions of ethics teaching programmes developed in the Ethics Education Programme described below. Since the launch of GEObs, the number of data entered has gradually increased, allowing better assessment of available expertise in various Member States.

Efforts are now focused on constructing the fourth database, which will seek to provide information about legislation, guidelines and policies Member States have developed in relation to the ethics of S&T. It will not merely provide the texts of such legal regulations but, first and foremost, identify the structure, set-up and contents that will be instructive to other countries that are contemplating the drafting of ethics legislation, for instance, in connection with research with human beings or the ethical principles of science in general. To provide useful information that can guide the drafting of legislation, it will provide examples; it will therefore be necessary to abstract or excerpt the main characteristics of existing legislation. A team of legal experts is currently examining the question of how international legislation can be made comparable; the team is developing a methodology for the construction of this database.

Ethics Education Programme

The Framework for Action drafted in 1999 at the World Conference on Science in Budapest states that ethics and the responsibility of science should be an integral part of the education and training of all scientists and that they should be encouraged to respect and adhere to basic ethical principles and responsibilities of science (World Conference on Science, 1999). In 2002, the Division of Ethics of Science and Technology and COMEST organized the Working Group on the Teaching of Ethics, which has provided advice on how to integrate ethics and responsibility in scientific training. This working group has produced a report on the teaching of ethics, which includes a survey of existing programmes, an analysis of their structure and contents, and detailed curriculum advice on how to integrate ethics, history, philosophy and the cultural impact of science into scientific education (COMEST, 2003b). This report has been the basis for the Ethics Education Programme launched in 2004.

During the 32nd UNESCO General Conference (2003), many Member States expressed the desire to initiate and support teaching programmes in ethics, not only in bioethics but in all scientific and professional education. Ethics teaching varies greatly between regions and countries, for it addresses moral issues of relevance to a given region. The first step is to collect data on ethics teaching. To establish a database of ethics teaching programmes, standardized forms have been developed to describe teaching programmes, so that the substance of each programme can be examined and various programmes analysed and compared. Within a group of countries, experts are identified who are actively teaching in a university setting. The experts are invited to take part in a regional meeting; in advance they are invited to provide data on their programmes and to return the forms so that these can be discussed during the meeting. Such meetings often provide experts with their first insights into the programmes in which their colleagues teach. At the meeting, data can be clarified and discussed, difficulties identified and discussed with colleagues. Once the empirical data has been obtained and clarified, the next step is to explore what will be necessary in the future, and how UNESCO can help to promote ethics teaching. Thus far, expert meetings have been held in Budapest (October 2004), Moscow (January 2005) and Split (November 2005). Approximately 100 teaching programmes have been validated and entered into the Global Ethics Observatory database. In 2006/7, further meetings are planned in Asia and the Arab region. One common finding so far is the vulnerability of ethics teaching programmes. While those teaching in such programmes are generally enthusiastic, the programme typically has no firm institutional basis, nor is there any systematic effort to create a future generation of ethics teachers. A pilot ethics teacher training course, organized by UNESCO and the UNESCO Chair in Bioethics in Haifa, Israel, has been set up to remedy this situation; it took place in November 2006 in Bucharest. Another finding is the lack of cooperation among nations. International cooperation of experienced teachers in neighbouring countries could give rise to programmes that have greater impact and sustainability, yet more to be done to increase teachers' awareness and willingness to work together in this area.

Another dimension of the Ethics Education Programme is the Advisory Expert Committee on the Teaching of Ethics. This ad hoc committee, which is composed of members of IBC and COMEST as well as representatives of the UNESCO Chairs in Bioethics, the Academy of Sciences for the Developing World (TWAS) and the World Medical Association (WMA), is assisting UNESCO in the area of ethics teaching. One of its first projects is to develop a proposal for a core curriculum in bioethics based on the recently adopted *Universal Declaration on Bioethics and Human Rights*. Once such a proposal has been drawn up, multimedia resources will be created to assist scholars who want to establish teaching programmes in bioethics in various cultures and regions. In the future, similar efforts can be made for other areas of applied ethics, such as environmental ethics, science ethics and engineering ethics (examples of which can be found in GEObs).

Assisting Bioethics Committees (ABC)

The *Universal Declaration on Bioethics and Human Rights* advocates the establishment of independent, multidisciplinary and pluralist ethics committees at national, regional, local or institutional levels. The purpose of these committees is to foster the exchange of ideas and information, support decision-making, develop tools for standard-setting, and strengthen coordination and contacts among experts and institutions—for example, through databases. They reinforce the role of UNESCO as an international clearing house for information on ethical issues. Moreover, ethics committees are intended as an important intermediary for the implementation of normative instruments adopted by Member States. While many countries have had experience with bioethics committees at various levels of government, the majority of Member States currently have no such committees. To achieve greater equity in this context, UNESCO has initiated a programme to support the establishment and operation of bioethics committees: the ABC (Assisting Bioethics Committees) programme. Through a series of practical guidebooks, information is provided about how to establish such committees and how they are to function once established (UNESCO, 2005a; 2006a). New guidebooks will address the topics of education of committee members and public outreach by committees. Task forces of experienced committee members in Member States with operational committees will assist countries that are in the process of establishing committees; they will also provide committee members with training in working procedures and ethical analysis. What type of ethics committee will be established in this process, whether a committee focused on bioethics or one with a wider scope (perhaps covering the whole range of ethics of S&T), will depend on the specific needs of the country in question.

AWARENESS-RAISING

UNESCO strives to foster better understanding of the major ethical issues raised by S&T, and supports analysis and discussion of those issues internationally, regionally and nationally. An essential part of this work is raising public awareness and stimulating public debate. This is important for two reasons. First, ethics is of interest to policy-makers because of public concerns. Due to public concern about and debate on issues such as cloning, research with human beings, transplantation, nuclear energy or environmental pollution and global warming, ethics has been placed on national and international agendas. Ethics is no longer solely the concern of scientists, engineers or healthcare professionals; it is no longer the exclusive province of scientific experts but now interests the general public as well. This development underscores the fact that science is first and foremost a public enterprise, a social activity and a cultural good. Second, scientific developments often affect or have the potential to affect all people. This is clear in medical research, which depends increasingly on the cooperation of large numbers of patients and healthy volunteers, often in international trials. Given the (potential) impact of science and research, the interests of scientists and researchers should be balanced with the research subjects' interests, precisely because the rights and freedoms of all citizens may be at stake in their work. Public debate and awareness-raising are therefore crucial if the general public is to achieve greater certainty that S&T are advancing within an ethical framework in which respect for human dignity and human rights is paramount. Scientists who are cognizant of their responsibilities to society take into account the effects their work may have on society and take precautions to assure that, for example, the environment is protected, justice promoted and biohazards and bio-events prevented.

That is why the Division of Ethics of Science and Technology of UNESCO is organizing 'Ethics around the World', a series of thematic rotating conferences designed to disseminate information and promote interaction and networking among national and international experts. In stimulating debate at national and regional levels, the division aims to increase participation of civil society in the debate. These conferences, which are jointly organized with national UNESCO commissions, UNESCO field offices and academic or research centres, usually feature one or two keynote speakers (often members of IBC or COMEST). Subsequent analysis and debate focus on specific topics relevant to the country in which the conference takes place. Over the last two years, 'Ethics around the World' conferences were held in the Netherlands, the Islamic Republic of Iran, Lithuania, Mexico, Argentina, the Russian

Federation, Portugal, Turkey, the Republic of Korea, Indonesia, China, Estonia, the Philippines, New Zealand and Peru.

Awareness-raising will also be stimulated by producing and disseminating publications. An explanatory brochure on ethics and human cloning has been published in the six official languages of UNESCO (Arabic, Chinese, English, French, Russian and Spanish) (UNESCO, 2005b). A similar brochure focusing on ethics and nanotechnology has just appeared (UNESCO, 2006b).

Another way to focus attention on ethics is the Avicenna Prize for Ethics in Science. Created in 2002 by UNESCO on the initiative of the Islamic Republic of Iran, this biennial prize rewards individuals and groups who have contributed to high-quality research in the field of ethics in S&T. Named for one of the greatest scientists, philosophers and doctors of the tenth and eleventh centuries, Abu Ali al-Husain ibn Abdallah ibn Sina (also known by his Latin name Avicenna; UNESCO, 2004), the prize consists of a gold medal of Avicenna, a sum of US$10,000 and a one-week visit to Iran, during which the prize-winner will deliver speeches at academic gatherings. Candidates are nominated by UNESCO Member States and international NGOs officially linked to the Organization, and the Director-General designates the winner on the recommendation of an independent international jury. The recipients of the Avicenna Prize have thus far been Margaret A. Somerville (2004) and Abdallah Daar (2006).

ETHICS AND EMERGING TECHNOLOGIES

Technological advances over the past centuries have substantially altered the human condition. For example, modern medicine and technology have become so intertwined that the physician has become the prototype of technological man. At the same time, it seems that technology and its concomitant changes in healthcare, particularly in the relationship between physician and patient, are also the source of many ethical worries about present-day healthcare. The increasing number of public debates about potential and actual uses of technology, as well as the differing views and values concerning human existence implied in judgements about innovations in S&T, have given rise to bioethics as a discipline and a subject of public and political discourse. For similar reasons, UNESCO has started a programme of activities in bioethics and ethics of S&T with the establishment of IBC and COMEST.

ISBN 978-92-3-104051-1

Since the 1970s, the need to evaluate new technologies has given rise to the new area of technology assessment (TA). A major impetus for establishing assessment agencies and procedures in many countries has been concern about the increasing costs of technologies. From the beginning, it was clear that TA is not based merely on technical and economic expertise, but that it also has to identify operative normative assumptions and thereby make explicit what previously was only implicit, for instance, in priority-setting and health policy. On the other hand, in daily practice it is obvious that ethics plays only a minor role in TA studies. This is unfortunate since TA can help to address ethical issues that are likely to arise through the development of new technologies. Rather than waiting until such issues emerge, it would be far more beneficial for policymaking and scientific development if a prospective approach were taken which aimed at the early identification and discussion of potential ethical issues. TA seems to oscillate between two poles: a narrow conception, which focuses on effectiveness, safety and economic impact of technologies, and a broad conception, which takes into account the social and ethical consequences of technologies. The narrow conception dominates current practices, although many experts advocate a more comprehensive approach. The question is how the role of ethics in TA can be better articulated.

If ethics is indeed intrinsically linked to TA, then it is high time to rethink the current concept of healthcare TA, transforming it into more comprehensive evaluation research (returning to the earlier, broader notion of TA). Such TA would not confine itself to issues of safety, efficiency and cost-effectiveness, but also would examine the value judgements at play in recommendations and determine if and how those recommendations were not simply scientific but also normative. Yet prior to reconsidering the concept of TA, it is important to look into why ethics has to date often been absent from TA studies. On closer inspection, at least two reasons may be discerned for why ethical analysis has rarely been incorporated in TA studies: first, technology is conceptualized in present-day evaluation research in such a way that it is demarcated from ethical issues. Second, applied ethics is itself often regarded as a variety of technology, one aimed at resolving, or at least 'pacifying', the moral consequences of the use of medical technologies.

In the first case, whether and where a line is drawn between ethics and TA depends on whether technology is regarded as value-neutral or value-laden. If it is held to be value-neutral, then technology clearly can be neither good nor bad, but is merely a means to some end. This seems to be the dominant view today in TA. Thus, for instance, it is common to differentiate between indirect and direct effects of technology, or TA in a narrow and a broad sense. The moral dimensions of new technologies were once considered largely to be 'second-order consequences', arising at the policy level when the data of evaluation studies were to be implemented in healthcare practice. This conception of TA assumes that ethics does not belong to the core processes of assessment, therefore that values are not intrinsically connected with technology itself but are related only to its application.

As for the second reason, the difficult relationship between ethics and TA is bound up with the nature of ethics itself. Bioethics in particular has developed into an autonomous discipline intended to aid healthcare practices, but it has also become part of the technological order. Often it is dominated by an engineering model of moral reasoning, relying on the idea of technological rationality when it addresses a given set of practical problems and applying a limited set of moral principles (particularly the principle of respect for autonomy). For this approach, bioethics is a sophisticated technology that can be employed to render a certain set of (potential) problems manageable and controllable. Technology confronts us with moral problems, according to philosophers such as Jürgen Habermas, Michel Foucault and Ivan Illich, basically because S&T penetrates, dominates and even 'colonizes' our lives and world. The solutions to such problems cannot come from an ethics that is itself technologically oriented; for when applied ethics approaches moral problems in a manner akin, say, to engineering – hence technically applying principles to cases and dilemmas – it itself proves to be but another manifestation of the same basic problem. It thereby becomes entangled in a vicious circle of sorts.

A repositioning of ethics will be necessary to uncover and analyse the moral dimension of practices of developing, testing and using technologies in present-day society. To this end, we note two ways in which ethics can make positive contributions in connection with TA.

The first concerns a category of moral questions that arise *within* the framework of a particular technology. Examples are debates on the moral status of the embryo in stem cell technology (examining whether the fertilized or 'activated' egg is similar to or different from the 'traditional' embryo), or on the conditions for gamete donation.

Questions of this type remain within the framework of the technology; they proceed from the acceptance of technology in its current form and seek to define its responsible and appropriate use. This type of question is usually addressed in TA studies – at least in those studies that include any ethical analysis. The theoretical framework for these ethical studies is provided by the present-day conception of 'applied ethics', that is, as the application of general ethical theories, principles and rules to specific problems that may arise in healthcare delivery, research and therapeutic practice. Such studies aim to analyse these problems and to offer morally justified solutions. The main instrument of this approach consists in a set of moral principles, usually three or four: respect for autonomy, beneficence, non-maleficence (which is sometimes included under beneficence), and justice. By comparison, the *Universal Declaration on Bioethics and Human Rights*, adopted in 2005 by the Member States of UNESCO, is far more comprehensive, consisting of fifteen principles in all. As normative generalizations, principles provide a basis from which to derive ethical guidelines and rules for human action, as well as the means by which to justify particular prescriptions and evaluations of such action. The advantage of principles is that they are defensible from a variety of theoretical, moral perspectives. They provide an analytical framework, a universal tool, for clarifying and resolving moral issues that arise in connection with a particular technology, something that TA alone cannot do.

Furthermore, ethics can contribute to TA by going *beyond* the framework of the technology itself. It then concentrates on a second category of moral questions about the technology itself. Here analysis focuses on the question of whether a technology as such is justified in the light of moral values. Such ethical analysis does not take the technology for granted, but starts from a critical perspective, assuming that technologies are not value-neutral but rather value-laden. Technologies are expressions of values, such as the values of searching for knowledge, having offspring or relieving suffering. However, these values are often implicit, not explicit. Ethical research is now beginning to take them as the starting point for a debate on (other) motivating values in society. This type of research focuses on values underlying or embedded in the development of technology itself. For example, studies of this kind do simply assume that the progress of transplantation technologies is beneficial. Instead, they question the specific framing of notions (such as personal integrity, altruism, death and the body) associated with these emerging technologies. For instance, they critically examine the tacit notion of 'body ownership', where the moral principle of respect for autonomy, while facilitating organ donation, reiterates and reinforces the traditional dualistic notion of the human person as an autonomous subject who possesses a material body. Such analysis also explores the recent expansion of these technologies through cell and gene transplantation. It calls attention to the claims of perfectibility ('enhancement') and immortality often implicit in the bewildering progress of stem cell technologies, and relates such claims to a philosophical, and sometimes utopian, body of knowledge. The methodology of such studies is historical as well as synthetic. They aim to provide both a diachronic and a synchronic perspective: values embodied in current technologies are explained in connection with similar values in history, but also in connection with developments in other scientific disciplines, thus looking beyond the framework of present-day disciplines. This type of ethics research assumes that ethics is first and foremost a philosophical endeavour to understand ourselves and our existence in terms of what is desirable or undesirable, supportable or reprehensible, good or bad. This approach to ethics can provide a rich perspective of relevant values that have to be taken into consideration in every responsible, political decision-making process and so makes a much needed contribution to TA.

The current gap between ethics and TA is remarkable since systematic assessment of technologies sprang from normative worries about the uncontrolled introduction of new technologies into different social contexts. In practice, TA has been confined to economic analysis. However, a broader approach to TA will be necessary in order to focus attention on ethical issues, to facilitate policy-making and to address social and ethical concerns in early stages of technological development. This prospective approach will also have to take account of the international dimension of TA. While new technologies tend to emerge in more developed countries, their impact is generally felt, sooner or later, in all countries, regardless of their stage of development.

NANOTECHNOLOGIES

Nanotechnology currently is one of the most rapidly developing fields of technology. It has many promising applications, for example in medicine, manufacturing and communication. Enormous benefits are expected in diverse areas, including drug development, information and communication technology (ICT), water decontamination, and production of stronger and lighter materials. Many governments and businesses are rapidly investing huge amounts of money in nanotechnologies. At the same time, however, concerns have been expressed about current and future developments. The impact of new nanomaterials on human health and the environment is uncertain, as are the type and scope of applications of the new technologies, and it is currently a matter of debate whether the regulatory frameworks now in place are adequate to deal with the new developments.

The term 'nanotechnology' refers to many different things. Searching the internet with the keywords 'nanotechnology' and 'definition' produced 1,850,000 hits as of October 2006. The prefix 'nano-' is derived from the Greek word for 'dwarf'. It refers to a scale of physical length: one nanometre (nm) is equal to one-billionth of a metre. A single human hair is about 80,000 nm in width; a red blood cell is approximately 7,000 nm wide; atoms are below a nanometre in size. It was Norio Taniguchi of Tokyo Science University who coined the term 'nanotechnology' in 1974 to designate a production technology at the nanometre level. It consists of the processes of separation, consolidation and deformation of materials by one atom or one molecule. The nanoscale therefore lies between the subatomic world and the classical physical world of biological cells, bacteria and viruses. At this scale, materials can have different or enhanced properties compared to the same materials at a larger size. In the US, the 21st Century Nanotechnology Research and Development Act (US Congress, 2003) defines nanotechnology as follows:

> The term 'nanotechnology' means the science and technology that will enable one to understand, measure, manipulate, and manufacture at the atomic, molecular, and supramolecular levels, aimed at creating materials, devices, and systems with fundamentally new molecular organization, properties, and functions.

In *Nanoscience and Nanotechnologies: Opportunities and Uncertainties*, the 2004 report of the UK's Royal Society and Royal Academy of Engineering, the distinction is made between 'nanoscience' and 'nanotechnology'. It defines nanoscience as 'the study of phenomena and manipulation of materials at atomic, molecular and macromolecular scales, where properties differ significantly from those at a larger scale'. Nanotechnology, on the other hand, is a term used for such a wide variety of tools, techniques and potential applications that it is more appropriate to refer to nanotechnologies in the plural. Nanotechnologies are defined as 'the design, characterization, production and application of structures, devices and systems by controlling shape and size at nanometre scale.' In the present volume, we shall follow the lead of that report and speak of 'nanotechnologies' in the plural.

As noted in a previous UNESCO publication (2006b), nanotechnologies are giving rise to various ethical questions. Like previous developments in S&T (e.g. nuclear energy, biotechnology and embryonic stem cells), the new technologies may generate intense public debate. In this case, however, since the technological developments are still in their early stages, we find that we have a rare opportunity to engage in prospective ethical reflection. Some have expressed doubts about whether new or unique ethical issues will actually arise in connection with nanotechnologies, while others have argued that a new area of applied ethics – 'nano-ethics' – is now emerging. Regardless of one's stance on such matters, what is most needed at this juncture is the recognition that there are indeed ethical issues – new or not – that require our attention now (Khushf, 2004). Future developments in nanotechnologies alone will enable us to determine whether the ethical issues to which they may give rise are actually new and whether the talk of a new sub-discipline of nano-ethics is justified.

Relevant here is the distinction made in the previous section concerning the scope of ethical questions in relation to new technologies. First there is the category of moral questions arising *within* the framework of particular technologies. They include questions of the possible adverse impact of nanotechnologies on health, safety and the environment. The Organisation for Economic Co-operation and Development (OECD), for example, has been working for some time to coordinate methods of assessing the affects of manufactured nanomaterials on human health and environmental safety. The second category of moral questions bears on the technologies themselves, relating the new technologies to ethical values in general. An important question here is the long-term impact of nanotechnologies on global society (Kearnes et al., 2006; Stone and Wolfe, 2006). It is unclear who will benefit and who might lose out due to these developments; there is a danger that nanotechnologies will intensify the gap between rich and poor countries, transforming it into a 'nano-divide'. Such questions go beyond concerns for risk assessment; they focus attention on global justice, science ethics and the goals of innovation in S&T (UNESCO, 2006b).

NANOTECHNOLOGIES AND UNESCO

Following the publication of *The Ethics and Politics of Nanotechnology* (UNESCO, 2006b), COMEST has been exploring the role that UNESCO, as an intergovernmental organization, can play in the ethical assessment of these emerging technologies. The brochure represents UNESCO's first effort to provide objective and comprehensive information on the ethical dimensions of nanotechnology. Without taking any position, the brochure aims to provide all Member States with an explanation of what nanotechnology is and to indicate a number of ethical issues raised by its development and applications. (The text is currently available only in English; translations into the other UN languages are in preparation.)

In 2005, an ad hoc group of experts (from Brazil, Canada, China, Germany, Japan, New Zealand, the Netherlands and the Republic of Korea) was established to study and analyse the ethical aspects of nanotechnology. The expert group met in UNESCO headquarters in Paris on 5–6 July and 6–7 December 2005. The group followed a twofold working strategy. The first phase involved the preparation of a 'state of the art' study on ethics and nanotechnology. This present volume is the outcome of that phase of work. The second phase consisted in producing materials and proposals for a UNESCO policy document outlining the kinds of international action that could be undertaken. The draft document proposes four types of action: awareness-raising, education, research and policy. This document is now under review by COMEST and was subject to further consultation and deliberation at the 5th Ordinary Session of COMEST, held in Dakar, Senegal in December 2006.

At the international level it is important to determine which policies are necessary. It must be asked, for example, whether and how the development of nanotechnologies can be steered towards the achievement of wider social and environmental goals? The answers to this question could be important for the entire international community. Another question is how governments can influence the possible 'nano-divide'. Many governments are now considering these and related questions, and many are working to develop (national or regional) policies. Recommendations have been made in reports and by scientists, policy-makers and advisory bodies (Royal Society & Royal Academy of Engineers, 2004; Foster, 2006). The recommendations include the following:

- Interdisciplinary research programmes should investigate the social and ethical issues expected to arise from the development of nanotechnologies.
- Formal training of all current and future researchers should include reflection on the ethical and social implications of nanotechnologies.
- The general public should be made more aware of the implications of these technologies and should be involved in debating them.
- Regulatory bodies and advisory committees should develop mechanisms to monitor and review the development of these technologies (particularly health, safety and environmental hazards), make the results of their reviews publicly available, and address any regulatory gaps.

Ethics may contribute to the evaluation of medical technologies in two ways. First, by mapping out the relevant moral issues arising within the framework of particular technologies. Second, going beyond this framework, recasting how problems are defined, exploring the interrelations of technical and non-technical issues, and examining the framework within which technology is analysed, that is, seeking to make explicit the implicit assumptions that often guide such analysis, not least of which is the priority given to economic concerns and the valorisation of instrumental rationality, which is central to the developed world.

UNESCO can play a decisive role by promoting and facilitating the exchange of information on new technologies among countries, as well as monitoring the assessment of these technologies to make sure that it adequately addresses values that are fundamental to particular countries and regions, while also respecting the values of the community of nations as a whole. Exploring the ethics of nanotechnologies accords with UNESCO's ethical mandate. First of all, ethical issues regarding nanotechnologies should be identified and analysed so that the general public, specialised groups and decision-makers can be made aware of the implications of the new technologies. Since nanotechnologies are developing quickly, a prospective or anticipatory approach to ethical issues is required. Instead of waiting for public concerns and moral discussions to emerge, international committees, such as the International Bioethics Committee (IBC) and the World Commission on the Ethics of Scientific Knowledge and Technology (COMEST),

NANOTECHNOLOGIES, ETHICS AND POLITICS – ISBN 978-92-3-104051-1 – © UNESCO 2007

must continuously monitor the possible benefits and harms of new and emerging technologies. Here, too, UNESCO can play a significant role by promoting, from a global perspective and at an international level, dialogue among all stakeholders and making recommendations to the decision-makers who will be challenged by the ethical issues of evolving and emerging technologies.

THIS VOLUME

The present volume is an outcome of the work of the aforementioned group of UNESCO experts. The group was asked to address two questions: (a) What is the state of the art of nanotechnology and what are the ethical issues related to it? (b) What opportunities are there for international action regarding these issues? Draft papers have been discussed and reviewed in the meetings. The book provides information and analysis in three main areas. Part 1 provides a review of the current state of the art in the scientific domain. Spangler (Chapter 1) and Liu (Chapter 2) explain what nanotechnologies are and give several examples of current research, various developmental trends and possible practical applications. Part 2 addresses ethical issues. Schummer (Chapter 3) provides a broad survey of ethical questions that nanotechnologies could potentially raise, whereas Gordijn (Chapter 4) and Evans (Chapter 5) focus on ethical problems in specific areas of medicine and health care. Part 3 treats policy. Court et al. (Chapter 6) explain the relevance of nanotechnologies for developing countries, and Choi (Chapter 7) shows how the general public can engage in debate on nanotechnologies and how educational programmes should be modified to increase sensitivity to moral issues among (future) scientists and researchers. By describing the experiences of the Commission on Ethics in Science and Technology (CEST) in Quebec, Canada, Jean et al. (Chapter 8) illustrate how policy-makers can prospectively address emerging technologies such as nanotechnologies.

BIBLIOGRAPHY

Borry, P., Schotsmans, P. and Dierickx, K. 2005. Developing countries and bioethical research. *New England Journal of Medicine*, Vol. 353, pp. 852–53.

COMEST. 2003a. *Proceedings Third Session*, 1–4 December 2003, Rio de Janeiro, Brazil. Paris, UNESCO.

———. 2003b. *The Teaching of Ethics*. Paris, UNESCO.
———. 2004. *Best Ethical Practice in Water Use*. Paris, UNESCO.
———. 2005a. *The Precautionary Principle*. Paris, UNESCO.
———. 2005b. *Proceedings Fourth Session*, 23–25 March 2005, Bangkok, Thailand. Paris, UNESCO.
Foster, L. E. (ed.). 2006. *Nanotechnology: Science, Innovation and Opportunity*. Upper Saddle River, NJ, Prentice Hall.
Kearnes, M., Macnaghten, P. and Wilsdon, J. 2006. *Governing at the Nanoscale: People, Policies and Emerging Technologies*. London, DEMOS.
Khushf, G. 2004. The ethics of nanotechnology: Vision and values for a new generation of science and engineering. National Academy of Engineering, ed., *Emerging Technologies and Ethical Issues in Engineering*. Washington, DC, The National Academies Press, pp. 28–55.
Lenoir, N. 1997. Are attitudes towards bioethics entering a new era? *Journal of Medical Ethics*, Vol. 23, pp. 69–70.
Pompidou, A. 2000. *The Ethics of Space Policy*. Paris, UNESCO.
The Royal Society & The Royal Academy of Engineering. 2004. *Nanoscience and Nanotechnologies: Opportunities and Uncertainties*. 29 July. London, The Royal Society & The Royal Academy of Engineering. http://www.nanotec.org.uk/finalReport.htm (Accessed 22 December 2006.)
Sass, H.-M. 1991. UNESCO Conference on Human Rights and Bioethics. *Kennedy Institute of Ethics Journal*, Vol. 1, pp. 253–56.
Stone, J. V. and Wolfe, A. K. 2006. Nanotechnology in society: Atlas in wonderland? *Practising Anthropology*, Vol. 28, No. 2, pp. 1–40.
UNESCO. 1945. *Constitution of UNESCO*. Paris, UNESCO. http://www.icomos.org/unesco/unesco_constitution.html (Accessed 22 December 2006.)
———. 1989. *25 C/Resolution 5.2*. Decision of the 25th session of the General Conference of UNESCO regarding 'Philosophy, ethics and the life sciences'.
———. 1991. *Science-Agenda Framework of Action*. World Conference on Science. Budapest, UNESCO.
———. 1993. *27 C/Resolution 5.15*. Decision of the 27th session of the General Conference of UNESCO regarding 'Preparation

of an international instrument for the protection of the human genome'.

———. 2004. *Avicenna and the Ethics of Science and Technology Today.* Paris, UNESCO.

———. 2005a. *Guide 1: Establishing Bioethics Committees.* Paris, UNESCO.

———. 2005b. *Human Cloning: Ethical issues.* Paris, UNESCO.

———. 2006a. *Guide 2: Bioethics Committees at Work: Procedures and Policies.* Paris, UNESCO.

———. 2006b. *The Ethics and Politics of Nanotechnology.* Paris, UNESCO.

———. 2006c. *Universal Declaration on Bioethics and Human Rights.* Paris, UNESCO. http://portal.unesco.org/shs/en/file_download.php/46133e1f4691e4c6e57566763d474a4dBioethics Declaration_EN.pdf (Accessed 16 October 2006.)

US Congress. 2003. *21st Century Nanotechnology Research and Development Act.* Public Law No. 108-153. Washington, DC. http://www.theorator.com/bills108/s189.html (Accessed 16 October 2006.)

Chapter 1
EVOLUTION OF NANOTECHNOLOGIES

Margareth Spangler Andrade

Research in nanoscale technologies is growing rapidly worldwide. Despite a lack of strategic focus, several subfields are expanding our ability to build structures with atomic precision. Public opinion is already divided between great expectations of future applications and fear of harmful consequences to the environment and to humankind that this new technology may bring.

This chapter presents aspects of the evolution of nanotechnologies thus far. Topics to be analysed include when and why nanoscience and nanotechnology captured the interest of researchers, how the field is evolving, its multidisciplinary character, the role scientists may play in its development, and the consequences of nanotechnology research and its applications in developing countries. The objective here is to stimulate discussion of questions that may arise and be answered at the present stage concerning ethics and nanotechnologies.

THE NANOWORLD

While nanotechnology is still in its infancy and controversy exists about its definition, which has changed over time, in this chapter we generally follow the definition adopted in this book's introduction, where nanotechnology is said to be 'the design, characterization, production and application of structures, devices and systems by controlling shape and size at nanometre scale' (Royal Society & Royal Academy of Engineering, 2004). This definition has the advantage encompassing all the key stages in the development of the concept and the field of nanotechnology.

While Richard P. Feynman made no mention of either 'nano' or 'nanotechnology' in his famous 1959 lecture at the California Institute of Technology (Caltech) (Feynman, 1960), the concept of nanotechnology was nevertheless operative in it. There he argued in detail for the possibility of putting together tiny things and even moving atoms or molecules one by one to create materials with different and striking properties, to build computers that could work at much higher velocities, and eventually to make incredible advances in the field of health. 'What I want to talk about', he said, 'is the problem of manipulating and controlling things on a small scale' (Feynman, 1960).

While Feynman's lecture may justly be argued to have inaugurated research in nanotechnology, there were in fact relevant discoveries that preceded his talk, starting in the nineteenth century, although they did not lead directly to nanotechnology (see timeline below). On the other hand, the end of the inaugural phase of the new technology may be said to have occurred with Binnig and Rohrer's invention of the Scanning Tunneling Microscope (STM) in 1981, after which the pace of new discoveries has accelerated immensely.

Michael Faraday discovers colloid gold		Langmuir discovers layers of one molecule		Binning and Rohrer invent the Scanning Tunneling Microscope	
1857		1932		1981	
	1905		1960		1985
	Albert Einstein explains the existence of colloids		Feynman suggests a world at nanoscale		A new form of carbon – the fullerene – is discovered

In the 1970s, scientists were dedicated to the study of growing thin layers to produce electronic and optical devices (Esaki and Tsu, 1970). They were interested in the creation of quantum wells, which are applied in opto-electronic devices such as lasers and photo-detectors.

Though still not using the term 'nanotechnology', these were the first one-dimensional nanostructures produced by scientists. The technique used was advanced epitaxy, the crystalline oriented deposition of layers of different materials with a thickness of only a few angstroms (one angstrom = 0.0000000001 m). The evolution of this research led in the 1980s to quantum wires, which are bi-dimensional nanostructures, and in the 1990s to the quantum dots (Gonzalez et al., 2000), and more recently to three-dimensional quantum structures with amazing applications in many fields, from electronics and optics to medicine. Quantum dots are very small particles with interesting optical properties: they absorb normal white light and, depending on their size, emit a range of bright colours. This property arises directly from the very small size of the particle.

In fact, the first to use the word 'nanotechnology' was a Japanese scientist named Norio Taniguchi, who coined it in 1974 to designate the production of a structure at the nanoscale by removing material from it (Taniguchi, 1974; OED). Taniguchi's notion of nanotechnology involved top-down miniaturization and was 'regarded as the latest stage in mechanical engineering, which has pursued ever-tighter precision of manufacture and tolerances throughout its history' (Budworth, 1996, p. 13). Top-down miniaturisation is the usual technique of cutting away material until a completed component or product is obtained. It is also the technique used to produce micro-electronic chips.

While Feynman's lecture posed a challenge to researchers in many fields, nanotechnology entered the popular imagination through Eric Drexler's futuristic book, *Engines of Creation* (1986). Inspired by Feynman's ideas, Drexler recovered the term 'nanotechnology' in the mid-1980s and gave it its current meaning. Because this was an important step towards the consolidation of the new scientific approach, Drexler is now generally considered to be the word's inventor. His concern lay with the advanced capabilities based on molecular assemblers, devices that would be able to guide chemical reactions by positioning reactive molecules with atomic precision. This is the bottom-up approach.

In 1985, the chemist R. E. Smalley and colleagues discovered a new arrangement of carbon atoms, which they named the 'fullerene' in honour of R. Buckminster Fuller and which later would also be called a 'buckyball' or 'buckyminster'. Its properties are completely different from the other two known forms of carbon: diamond and graphite. The fullerene is a molecule that consists of 60 carbon atoms (Figure 1.1).

Figure 1.1. Fullerene structure of carbon atoms

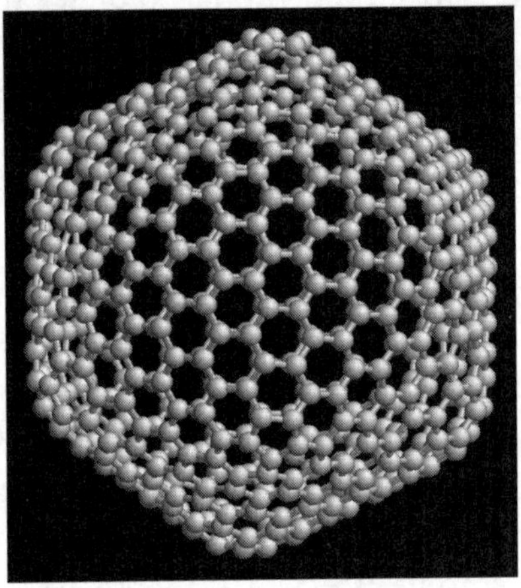

Source: http://en.wikipedia.org/wiki/Fullerene.

The discovery of the fullerene helped to put the then-emerging field of nanotechnology, which involves making products from designer molecules, into the limelight. The subsequent development of Smalley's research (Wang et al., 2005) led to the tubular variant of the fullerene, the carbon nanotubes (Figure 1.2). When made with molecular perfection, these tubular fullerenes offer revolutionary electrical, thermal and mechanical properties on the nanometre scale. Researchers believe that major new technologies will be developed over the coming decades from fullerene tubes, fibres and cables. These nanotubes are also envisaged to be used to transport drugs in the human body or to strengthen materials for aircraft parts. Nowadays many groups across

Figure 1.2. The carbon nanotube

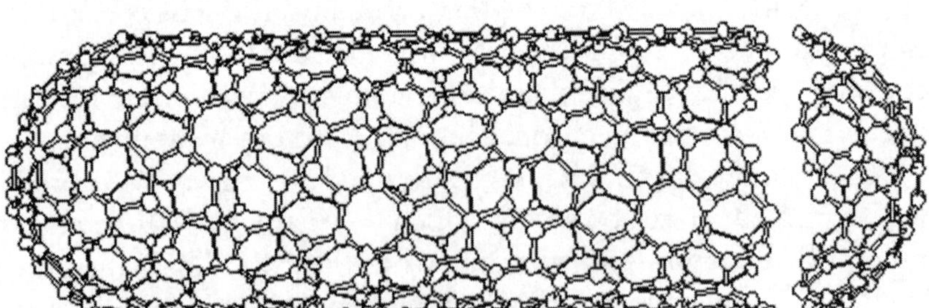

Source: R. Saito. http://www.godunov.com/Bucky/nanotube.html
Copyright © Tohoku University

the globe are dedicated to the development of the underlying basic science of nanotubes, as well as of methods of producing, purifying, characterizing and assembling them.

The well-known debates between Drexler and Smalley over the possibilities of technological applications in the nanoworld (Smalley, 2001) centred on the possibility of one day creating self-replicating nanomachines. Despite their differences, both scientists agreed that nanotechnology would most likely result in enormous advances in many areas of human life.

The fictional work of two other authors, based on Drexler's ideas, has also influenced scientists and public opinion. Erik Viktor, a talented Belgian artist-writer, has produced fantastic drawings of nanorobots, or nanobots, acting in the blood stream to run tests and diagnostics or to facilitate healing (Viktor, 2002). Michael Crichton's bestseller *Prey* introduced the fear that self-replicating nanobots could go out of control (Crichton, 2002). In such works, the stage has been set for both great expectations and great fear of the unknown and the uncontrollable course nanotechnology could take.

SCANNING PROBE MICROSCOPES

Another leap forward in the development of nanotechnology came in the 1980s, prior to the discovery of fullerenes, with the invention of a new family of microscopes, the Scanning Probe Microscope (SPM). The researchers at IBM-Zurich showed that it was possible finally to 'see' and access atoms one by one. The patriarch of this microscope family, the Scanning Tunneling Microscope (STM) devised by Gerd Binnig and Heinrich Rohrer, became known when Binnig, Rohrer, Gerber, and Wiebel published two now-classic papers in the *Physical Review Letters* (1982, 1983), which included an image of a repeating pattern of atoms on what is known as the silicon 7-by-7 surface (Binnig et al., 1982).

SPMs do not use lenses to produce the magnified image. Instead, a 'local' probe is scanned over the surface of the specimen and measures some physical property associated with the surface. This probe is produced from a material appropriate for the measurement of the particular surface property (Miles et al., 2003). The scanning process is simply mechanical, but with extremely high precision. The main advantages of SPM are that it can produce three-dimensional images of a surface up to atomic resolution, and that it offers the possibility operating in different environments, such as in liquid, vacuum and gas. Also sample preparation is considerably less time-consuming and invasive than for conventional microscopes. These capabilities are of importance for imaging biological systems in physiologically relevant environments and for the study of the crystallization, dissolution or corrosion of materials in various solvents.

It is remarkable that already in the late 1950s Feynman saw the necessity of developing new microscopes capable of achieving higher levels of magnification. 'Make the microscope one hundred times more powerful', he said, 'and many problems of biology would be made very much easier'; and he concluded: 'I put this out as a challenge: Is there no way to make the electron microscope more powerful?' (Feynman, 1960).

As far back as 1957, electron diffraction had been used to probe the silicon 7-by-7 crystalline surface, but the results were so vague that they were consistent with several different models for the arrangement of atoms on the surface. The STM image almost single-handedly settled this issue, dramatically illustrating the power of the new tool (Binnig et al., 1982). That image of silicon was a revolution.

It is interesting to recall that in the first years that new invention encountered pervasive disbelief and even some hostility. Given previous experience with phenomena on an atomic scale, especially the Quantum Mechanics Uncertainty Principle, the possibility of atomic resolution seemed remote if not altogether theoretically impossible. Many people at that time believed that those images had no meaning at all, aside from computer simulation. Nowadays, with new devices and consistent work, the battle against disbelief has been won. Scientists still have to separate artefact from reality, but they no longer doubt it can be done. Interpretations are still much disputed, but SPMs are becoming part of the common furniture in different types of laboratory.

SPMs can map such properties as magnetism, surface roughness and electrical potential, depending on the kind of probe they are equipped with. They even allow scientists to reach into the nanoworld and change it, turning nanoscience into nanotechnology. SPMs can touch things and even hold them. They can manipulate atoms one by one.

Probe selection depends on the operation mode and the sample type. An increasing number of SPMs have been developed and commercialized since the 1990s. Today researchers can get to the heart of the matter in the nanoworld, monitoring individual atoms or molecules and observing how their particular environments affect them. This can be performed for different kinds of materials, including virtually non-prepared biological material. As a result, nanomanipulation and nanolithography have undergone consistent improvement.

It is interesting to note the feedback of the nanotechnology discoveries on new nano-applications. For example, carbon nanotubes are used nowadays mounted on the probes of SPMs to obtain finer tips and consequently better images (Figure 1.3). With better, high resolution images, scientists can go deep into the material and produce new knowledge and applications. In fact, it is expected that the advances of nanotechnology will have an exponential effect on the development of nanotechnology itself.

Figure 1.3. Carbon nanotube SPM tip probe

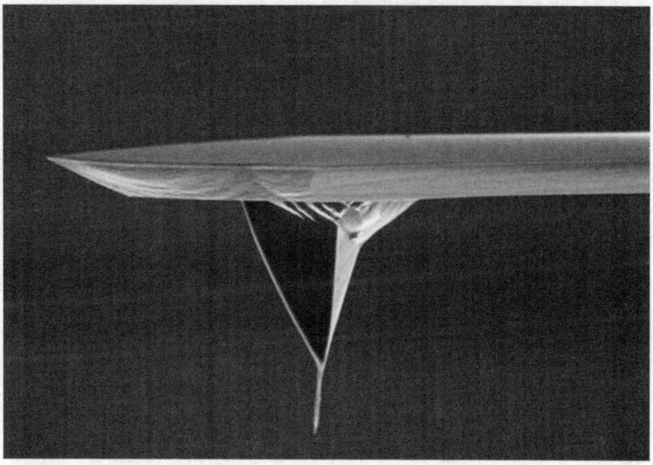

Source: http://www.xintek.com/products/cathodes/AFM_tip.html
Copyright © Xintek Inc

The SPM technique has become so important for nanotechnologies that meetings and conferences are dedicated exclusively to it. The Third Latin-American Symposium on Scanning Probe Microscopy was held on 18–20 April 2005 in Ouro Preto, Brazil (III LASPM, 2005). It was attended by researchers from throughout Latin America, the United States, Europe and China, and six equipment exhibitors presented state-of-the-art SPMs for different types of materials and applications.

THE MULTIDISCIPLINARY CHARACTER OF NANOTECHNOLOGY

There have been many discussions about whether nanotechnology will make a technological breakthrough once multidisciplinary projects have been set up. Nowadays audiences at nanotechnology meetings are diverse, and interactions between different scientific areas have increased. Many efforts all over the world have been made to establish multidisciplinary groups for studying different subjects, including biology, physics, chemistry, health and computer science. Any researcher who has participated in such a team knows about the problems that must be overcome to achieve a fruitful integration of the various disciplines involved. Nonetheless, multidisciplinary work is inevitable when ethical issues are at stake. For instance, a carbon nanotube can be used as an envelope for targeted drug delivery, but is it safe from the point of view of toxicity or blood clot formation? And can the rejects of carbon nanotube production threaten living creatures in the way that, for example, asbestos does? These sorts of questions are best addressed in a multidisciplinary context, where it is possible to gather information and insights from different sectors of society so as to discuss the issue before it becomes critical.

Here too the SPM has influenced the trend towards multidisciplinarity in nanotechnology. One relevant feature of the SPM family is that it can deal with different materials relatively easily. Other types of microscopes demand specific knowledge of sample preparation and analysis of the images, which differ, for example, for materials scientists and biologists. In the SPMs, only surface properties of the sample are investigated. Materials scientists are well acquainted with this type of analysis and could help researchers from other areas in learning about the use of SPMs. This could enable groups composed of scientists from different areas (such as physics, chemistry, materials science and engineering, biology, pharmacy and medicine) to begin to work together. Through such groups, knowledge can spread more quickly, and environmental and health problems can be identified more easily.

SCIENTISTS AND NANOTECHNOLOGY

Scientists are often more concerned about understanding the behaviour of things in the universe than about the hidden implications of a new finding or the ethical issues involved therein. Generally, what is more important to scientists than the application of any new knowledge is the need to gather data that demonstrate the truth of their findings, and to do so in a diversity of ways, using varying techniques and approaches. While many may object to this characterization, it is nevertheless obvious that a division exists between scientific research and the assessment of ethical issues connected with the results of that research. Clearly it would not make sense to engage in ethical analysis at any and every point in the course of scientific research, but it is just as clear that a balance between the two (research and ethical analysis) must be struck which is in everyone's best interest.

Protocols imposed on pharmacy, medicine and health sciences for dealing with a new drug or procedure have shown themselves to be quite reliable; they should not be considered a hindrance to scientific work. While they must be followed, they can also be improved and adjusted to new situations wherever necessary, as has been done in the case of nanotechnologies. The scientific community is organized to do just that.

But is it also ready to cope with the multi- and interdisciplinary character of nanotechnologies? In fact, the medium- and the long-term consequences of these new technologies require greater attention. This in and of itself provides good reason to promote meetings of and discussions among experts from the exact sciences, applied disciplines, and even the humanities. Many worldwide initiatives have already been implemented to develop human resources through innovative nano-education programmes. Such programmes seek to reach young people and to foster the development of more flexible perspectives, disciplines, sectors and cultures. It is expected that nano-education can help to advance the growing international movement towards interdisciplinary science education (Chang, 2006).

It also must be borne in mind that the properties of the newly proposed nanomaterials are not yet well-known. These properties can be quite different from those of bulk material, that is, from the properties measured for a large amount of the material. In many cases, they cannot even be measured by the conventional methods and techniques used to study bulk material. Researchers will have to devise new methods of measurement and tests in order to access and take account of the precise properties in a nanoworld.

In this respect, metrology is a fundamental field that must be further developed. Not only are dimensional measurements important, but the hardness and elasticity of materials, for example, can also influence an application. The field of mechanical properties measurement on the nanoscale has received a great deal of attention, yet many unanswered questions remain (Oliver and Pharr, 1992). Metrology is also important to help in the exchange of scientific knowledge among groups and countries. It is a tool that enables everyone to speak the same language and assures that the advances of developed countries will be accessible to and assimilable by less developed ones. This is of great relevance in preventing the widening of the gap between nations and regions.

NANOTECHNOLOGY IN BRAZIL

Developing countries already have groups of scientists researching nanoscience and nanotechnologies. In South America, Argentina, Brazil, Chile and Mexico have investment programmes for nanotechnologies. Three virtual institutes, four national networks, and a large number of researchers are involved in nanotechnology in Brazil in an attempt to keep pace with world progress and to develop solutions for its most pressing problems, such as emergent viruses, malaria, chagas disease, dengue fever, cancer, degenerative diseases, energy and nano-opto-electronics.

The Brazilian Program on Nanotechnology has a projected budget of approximately US$40 million for 2004–2007, and aims to develop new products and processes in nanotechnology, as well as their industrial application. This amount can be compared with India's S&T budget for 2004–2009, which is approximately US$20 million. The budgets of developing countries are no doubt far from those of the US, Japan, China and South Korea, the leading nations in the field.

The Brazilian programme should yield results that will give the country's economy extra impetus regarding energy, petrochemistry, automobiles, IT, medicine, pharmacy, metallurgy, mining, environmental protection and electro-electronics. Moreover, similar impact is expected in strategic areas, such as national, personal, patrimonial and food security.

University and research institute laboratories in Brazil are well equipped for laboratory-scale production, analysis and characterization of nanoproducts. The number of specialized publications from these institutions is quite high. In addition, considerable progress has been achieved in scanning probe microscopy work in Latin America. Ten years ago, only four SPMs could be found in Brazil; today more than eighty of these instruments are operating in the country, and the quantity is increasing rapidly. Furthermore, this equipment is involved in useful work, not only on scientific grounds but also in industrial development and innovation programmes. Even so, the number of research results successfully translated into commercial applications has been quite low. This situation continues to be a nagging problem, one that must be solved if Brazil is to succeed in the world of nanotechnology.

Globally, roughly two million new nanoscientists and engineers will be needed in the next ten–fifteen years (Chang, 2006). In the meantime, it is becoming increasingly important to integrate education in research and prepare young researchers to collaborate across disciplines, sectors and cultures. Recognizing this, Brazil has initiated a special project for human resources development in nanotechnology which emphasizes undergraduate- and graduate-level courses. The number of researchers in many areas of nanotechnology is expected to at least double. Other goals of this programme are to increase the number of companies incorporating nanotechnology products or processes by 200 percent and to double deposited patents. Support for existing institutes and laboratories working in the field will be given in the form of new equipment and maintenance. Doctors are encouraged to travel abroad and work with important groups in different countries. Furthermore, classes in ethics and philosophy will be implemented in elementary and secondary schools.

7. CONCLUDING REMARKS

Four observations follow from this discussion:

1. The rapid evolution of nanotechnology in many areas requires the development of an organizational structure capable of promoting more interaction among experts from the sciences, engineering and the humanities.
2. Education, from the primary to the tertiary level, will play a crucial role in coming developments in S&T, particularly in nanotechnology. Its task in view of the rapid expansion of S&T will be to prepare individuals to engage in scientific research and engineering *and* to deal with developments in an ethical manner. As time goes on, more and more highly skilled researchers with broad educational backgrounds will be needed.

3. Metrology at the nanoscale promises to provide a means for avoiding the concentration of knowledge in rich countries. The development of new measurement techniques and the assessment of standard methodologies are needed in order to reach a higher, more exacting level in nanometrology.
4. Research projects involving groups from different countries should help to bring about a more uniform distribution of knowledge and to prevent the deepening of the economic and social gap already existing among more- and less-developed countries.

BIBLIOGRAPHY

Binnig, G., Rohrer, H., Gerber, C. and Wiebel, E. 1982. Surface studies by scanning tunneling microscopy. *Physical Review Letters*, Vol. 49, pp. 57–61.

———. 1983. 7 × 7 reconstruction on Si(111) resolved in real space. *Physical Review Letters*, Vol. 50, No. 2, pp. 120–23.

Budworth, D. W. 1996. *Overview of Activities on Nanotechnology and Related Technologies*. Seville, Institute for Prospective Technological Studies.

Chang, R. P. H. 2006. A call for nanoscience education. *Nano Today*, Vol. 1, No. 2, pp. 6–7.

Crichton, M. 2002. *Prey*. New York, HarperCollins.

Drexler, K. E. 1986. *Engines of Creation: The Coming Era of Nanotechnology*. New York, Anchor Books. http://www.e-drexler.com/d/06/00/EOC/EOC_Table_of_Contents.html (Accessed 16 December 2006.)

Esaki, L. and Tsu, R. 1970. Superlattice and negative differential conductivity in semiconductors. *IBM Journal of Research and Development*, Vol. 14, pp. 61–65.

Feynman, R. P. 1960. There's plenty of room at the bottom. *Engineering and Science*, Vol. 23, No. 5, pp. 22–26, 30, 34, 36. http://www.zyvex.com/nanotech/feynman.html (Accessed 20 December 2006.)

Gonzalez, J. C., Rodrigues, W. N., Moreira, M. V. B., Oliveira, A. G., da Silva, M. I. N., Vilela, J. M. C., Andrade, M. S., Ugarte, D. and Silva, P. C. 2000. On three-dimensional self-organization and optical properties of InAs quantum-dot multilayers. *Applied Physics Letters*, Vol. 76, No. 23, pp. 3400–2.

Wikipedia. n.d. Norio Taniguchi. http://en.wikipedia.org/wiki/Norio_Taniguchi (Accessed 16 December 2006.)

Miles, M., Antognozzi, M., Haschke, H., Hobbs, J., Humphris, A. and McMaster, T. 2003. Tour de force microscopy. *Materials Today*, Vol. 6, No. 2, pp. 30–37.

Oliver, W. C. and Pharr, G. M. 1992. An improved technique for determining hardness and elastic modulus using load and displacement sensing indentation experiment. *Journal of Materials Research*, Vol. 7, No. 6, pp. 1564–83.

Smalley, R. E. 2001. Of chemistry, love and nanobots. *Scientific American*, Vol. 285, No. 3, pp. 68–69.

Taniguchi, N. 1974. On the basic concept of nanotechnology. *Proceedings of the International Conference on Production Engineering*. Tokyo, Japan Society of Precision Engineering, pp. 18–13.

Third Latin American Symposium on Scanning Probe Microscopy (III LASPM). 2005. *Microscopy and Microanalysis*, Vol. 11, Supplement S03.

Viktor, E. 2002. *Spaceworld2000*. http://www.spaceworld2000.com/nanotechnology.html, or http://www.foresight.org/Nanomedicine/Gallery/Artist/Viktor.html (Accessed 16 December 2006.)

Wang, Y. K., Kim, M. J., Shan, H. W., Kittrell, C., Fan, H., Ericson, L., Hwang, W. F., Arepalli, S., Hauge, R. H. and Smalley, R. E. 2005. Continued growth of single-walled carbon nanotubes. *Nano Letters*, Vol. 5, pp. 997–1002.

Chapter 2

NANOTECHNOLOGY: THE STATE OF THE ART

Jixing Liu

Based on the definition and evolution of nanotechnology as addressed in the two previous chapters, this chapter will provide a survey of the state of art of nanotechnology. Since this rapidly growing field encompasses such wide-ranging areas – from basic branches of natural science (such as chemistry, physics, biology) to applied areas (such as medicine, engineering and electronics) – and is fuelled by substantial funding from governments and industries all over the world, an adequate account of the state of the art of nanotechnology is more than any one person can tackle alone. The scope of our undertaking in this chapter will therefore be restricted to only two objectives: (1) to give a brief overview of some key achievements in this field and highlight some recent advances in light of official, published reports and articles; and (2) to indicate the main trends in the field by way of examples rather than comprehensive general statements and technical explanations. In this connection, the recent report on nanotechnology by the Royal Society and the Royal Engineering Academy (2004) has been very helpful.[1]

If one disregards the progress made so far in diverse areas of nanotechnology, one might be inclined to agree with the following statements:

1 In preparation of this chapter, the contents of the comprehensive survey report, Royal Society & Royal Academy of Engineering, 2004, chapters 3 and 4, have been used extensively, at times without explicit mention.

1. The most notable achievements have occurred in the area of materials for which researchers have not only accumulated systematic knowledge of specific properties of nanomaterials but have also already had direct contact with real products made of such material.
2. The most puzzling achievements to date have been in nanoelectronics and information technology (IT). While many new exciting advances in those areas have been reported, no one knows yet when, for example, computers with the resultant capacity to function powerfully while being extremely small will be available.
3. The most encouraging achievements are in the area of medicine, where quite a few effective means to improve healthcare for human beings have been discovered, some of which are nearly ready for clinical use. There are some concerns, however, about the safety and long-term effects of these medical applications of nanotechnology.
4. The innovations in the manufacturing industry, compared to other areas, are rather slow. Since most innovations are still in the laboratory test stage, it will be some time before we will see significant progress in industrial applications of nanotechnology.
5. To meet the stringent requirements of the nanotechnology-based manufacturing industry, more innovative measurement techniques and instruments are urgently needed.

These statements, which together give one possible account of the state of the art of nanotechnology, will undergo substantial revision in what follows. In the next sections, we shall look more closely at nanomaterials, instrumentation and nanometrology, bio-nanotechnology and nanomedicine, and industry applications. Given that the aim of this chapter is to provide the scientific background necessary for more informed participation in the ethical discussion carried out in the last two parts of this book, our emphasis here will be on two topics from the foregoing list: *nanomaterials*, where many concerns about risks are raised, and *nanomedicine*, where the main focus is on the ethical debate. In each section, after briefly sketching the current state of development of the field, we present one or two of the most recent advances as illustrative examples.

NANOMATERIALS

As introduced in the previous chapter, nanomaterials are materials that have structured components with at least one dimension measuring less than 100 nm. Materials that have one nanoscale dimension are layers, such as films or surface coatings. Some parts of computer chips fall in this category. Materials that are nanoscale in two dimensions include nanowires and nanotubes. Materials that are nanoscale in three dimensions are particles, for example, precipitates, colloids, and tiny particles of semiconductor materials, or quantum dots. Nanocrystaline materials made of nanometre-sized grains also belong in this category. The specific properties of nanomaterials are determined by two principal factors: increased relative surface area, and quantum effects. These factors can change or enhance properties of nanomaterials, such as reactivity, strength, and electric characteristics. At the larger end of the size range of nanomaterials, other effects such as surface tension or 'stickiness' are important; these also affect physical and chemical properties. For liquid or gaseous environments, Brownian motion, which describes the random movement of larger particles or molecules owing to their bombardment by smaller molecules and atoms, is also important. This effect makes control of individual atoms or molecules in these environments extremely difficult.

One-dimensional nanomaterials, such as thin films and engineering surfaces, have been developed and used for decades in fields such as electronic device manufacturing, chemistry and engineering. In the silicon integrated-circuit industry, many devices rely on thin films for their operation, and the manipulation of film thickness approaching the atomic levels is routine. Monolayers (layers in depth of one atomic or molecule size) also are routine in chemistry. The formation and properties of these layers from the atomic level up are quite well understood. The advances currently being made are in the control of the composition and smoothness of surfaces, and the growth of films. Engineered surfaces with tailored properties, such as large surface area of specific reactivity, are now used routinely in a range of applications, including fuel cells and catalysts.

Basic building blocks of nanomaterials

The versatile building blocks of nanostructures, nanotubes, nanowires, fullerenes, quantum dots and dendrimers have generated considerable interest among the S&T communities in recent years. The novel electrical and mechanical properties of these materials stirred intense basic and applied research. Recently, biopolymers such as DNA and protein have frequently been used as components of hybrid nanostructures. This provides important opportunities for the fabrication of functional devices in many technical and industrial disciplines. In addition, different deliberately manufactured nanoparticles, as raw material for composites, play important roles in the application of nanotechnology. The main research results for these building blocks follow.

Carbon nanotubes

Carbon nanotubes (CNTs) were first observed in 1991. They are extended tubes of rolled graphite sheets. There are two kinds of CNTs: single-walled (one tube) and multi-walled (several concentric tubes). Both are typically a few nanometres in diametre and several micrometres to centimetres in length. They are very strong mechanically (as stiff as diamond), flexible (about the axis) and can conduct electricity extremely well, and under certain conditions the single-walled nanotube even shows superconductivity (Tang et al., 2001), or behaves as a semiconductor. The optical properties of single-walled semiconducting CNTs are currently the focus of intense study (Bachilo et al., 2002; Misewich et al., 2003; Wang et al., 2005; Chen et al., 2005). All these properties give them a wide range of potential applications, such as in reinforced composites, sensors, nanoelectronics and display devices.

Inorganic nanotubes

Soon after the discovery of CNTs, inorganic nanotubes and fullerene-like materials based on layered compounds, such as molybdenum disulphide, were discovered. They have excellent lubricating properties, resistance to shockwave impact, catalytic reactivity and high capacity for hydrogen and lithium storage, which suggests a range of promising applications. In particular, oxide-based nanotubes are being explored for their applications in catalysts, photo-catalysts and energy storage.

Nanowires

Nanowires are ultrafine wires or linear arrays of dots, formed by self-assembly. They can be made from a wide range of materials. Semiconductor nanowires made of silicon, gallium nitride and indium phosphide have demonstrated remarkable optical, electronic and magnetic characteristics (for example, silica nanowire can bend light around very tight corners). Nanowires have potential applications in high-density data storage: either as magnetic read-head or as patterned storage media, or as electronic and opto-electronic nanodevices for metallic interconnects of quantum devices and nanodevices. The preparation of these nanowires relies on sophisticated growth techniques, such as self-assembly processes, chemical vapour deposition and molecular beam epitaxy. How to control the growth conditions for producing nanowires with required properties remains one of the challenges. A very recent study has taken an important step closer towards high-volume production of nanowire of opto-electronic material cadmium selenide (Ma and Wang, 2005).

Biopolymers

The variability and site recognition of biopolymers, such as DNA molecules, offer a wide range of opportunities for the self-organization of wire nanostructures into much more complex patterns. The backbones of DNA may then, for example, be coated in metal. These materials also offer opportunities to link nanotechnology and biotechnology in, for example, biocompatible sensors and small, simple motors. The self-assembly of organic backbone nanostructures is often controlled by weak interactions, such as hydrogen bonds and hydrophobic, Van der Waals interactions (generally in aqueous environments). The combinatory nanostructures consisting of biopolymers and inorganic compounds open up a number of scientific and technological opportunities, and this area of research in nanotechnology has become very active.

Fullerenes (Carbon-60)

As was mentioned in the previous chapters, carbon-60 (C_{60}) is a new class of carbon material that was discovered in the mid-1980s. These are spherical molecules about 1 nm in diameter, comprising 60 carbon atoms arranged as 20 hexagons and 12 pentagons, thus it has the configuration of a soccer ball. In 1990, a technique to produce large quantities of C_{60} was developed by resistively heating graphite rods in a helium atmosphere. Several applications are envisioned for fullerenes, such as using them as miniature 'ball bearings' to lubricate surfaces, in drug delivery vehicles and in electronic circuits. A recent investigation suggests that subjecting fullerenes to high pressure can produce a new type of super-tough material, which is stronger than diamond (Dubrovinskaia et al., 2005).

Quantum dots

Nanoparticles of semiconductors (quantum dots) were theorized in the 1970s and first created in the early 1980s. If semiconductor particles are made small enough, quantum effects come into play which limit the energies at which electrons and holes (the absence of an electron) can exist in the particles. Since energy is related to wavelength (or colour), this means that the optical properties of the particle can be finely tuned, depending on its size. Thus particles can be made to emit or absorb specific wavelengths (colours) of light merely by controlling their size. Recently, applications for quantum dots have been found in composites, solar cells and fluorescent biological labels (for example, to trace a biological molecule). Advances in chemistry have resulted in the preparation of monolayer-protected, high-quality, mono-dispersed, crystalline quantum dots as small as 2 nm in diameter, which can be conveniently treated and processed as a typical chemical reagent.

Dendrimers

Dendrimers are spherical polymeric molecules formed through a nanoscale hierarchical self-assembly process. There are many types of dendrimers, the smallest being several nanometres in size. Dendrimers are used in conventional applications, such as coatings and inks, but they also have a range of interesting properties that could lead to useful applications. For example, dendrimers can act as nanoscale carrier molecules and as such could be used in drug delivery. Environmental clean-up could be assisted by dendrimers, since they can trap metal ions that could then be filtered out of water with ultra-filtration techniques.

Nanoparticles

Nanoparticles widely exist in nature: they are produced, for example, by photochemical and volcanic activity, and by plants and algae. They also have been created for thousands of years as by-products of combustion and food cooking, and more recently of vehicle exhausts. Deliberately manufactured nanoparticles, such as metal oxides, are far outweighed in quantity by those arising naturally and from pollutants. Nanoparticles are of interest due to the new properties they exhibit. For example, titanium dioxide and zinc oxide become transparent at nanoscale, though they can absorb and reflect ultraviolet light and so have possible applications in sunscreens. Nanoparticles have been used in new cosmetics, textiles and paints, as well as in methods of targeted drug delivery. These particles can also be arranged in layers on surfaces, providing a large surface and hence enhanced activity relevant to a range of potential applications, such as catalysts. Several recent studies have shown that the specific shape of nanoparticle aggregates could greatly change surface properties, which may prove useful in many contexts. One interesting example is a highly water-repellent surface that mimicked the dual-size roughness structure of the lotus leaf; the coating consisted of a raspberry-like particle aggregate made of silica spheres bonded to an epoxy-based polymer film (Figure2.1; Ming et al., 2005).

Figure 2.1. The surface topology (AFM phase image, right) of super-hydrophobic films containing silica-based particles that look similar to a raspberry (left)

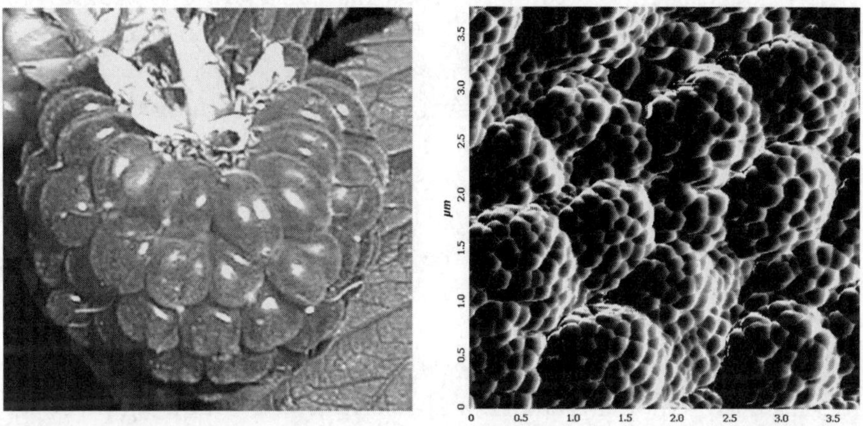

Source: Ming et al., 2005. Copyright © Eindhoven University of Technology

Key applications of nanomaterials

What are nanomaterials already used for?

To pay the bills of basic and applied research, nanoscale materials are already present in a wide range of products. Among the most well-known is a type of glass for windows which is coated with titanium oxide nanoparticles that react to sunlight and break down dirt. When water hits the glass, it spreads evenly over the surface, instead of forming droplets, and runs off rapidly, taking the dirt with it. Nanomaterials are used by the auto industry to reinforce certain properties of car bumpers and to improve the adhesive properties of paints. Other uses of nanomaterials in consumer products include:

- **Sunglasses** using protective and antireflective ultra-thin polymer coatings. Nanomaterials also offer scratch-resistant coatings based on transparent and ultra-thin nanocomposites.
- **Textiles** with special properties. The windproof and waterproof properties of one type of ski jacket, for example, are obtained not through a surface coating but through the use of nanofibres.
- **Sports equipment:** A high-performance ski wax, which produces a hard and fast-gliding surface, is already in use. The ultra-thin coating lasts longer than conventional waxing systems. Tennis rackets with carbon nanotubes have increased torsion and flex resistance. The rackets are more rigid than conventional carbon rackets. Coating the inner core of tennis balls with lacy polymer nanocomposites give them twice the lifetime of conventional balls.
- **Sunscreens and cosmetics** based on nanomaterials are already widely used. Examples are nanosized titanium dioxide in sunscreens to reflect UV light, and nanosized iron oxide in some lipsticks as a pigment.
- **Tougher and harder cutting tools** made of nanocrystalline materials, such as tungsten carbide, tantalum carbide and titanium carbide, are more wear- and erosion-resistant, and last longer than their conventional (large-grained) counterparts. They are finding applications in the drills used to bore holes in circuit boards.

Which future applications may nanomaterials have?

Research on nanomaterials strongly suggests the following practical applications in the foreseeable future:

- **Paints:** Incorporating nanoparticles in paints could improve their performance, for example by making them lighter and giving them different properties. Thinner paint coatings used on aircraft would reduce their weight, which could be beneficial to the environment. Anti-fouling surface treatment is also valuable in process applications such as heat exchange, where it could lead to energy savings. Some of the nanoparticle paints could act as distributed sources of electrical power. Other applications for nanoparticles might lie in paints that change colour in response to changes in temperature or chemical environment, or paints that have reduced infrared absorption and so reduce heat loss.

- **Remediation:** The potential of nanoparticles to react with pollutants in soil and groundwater and to transform them into harmless compounds is currently being investigated. In one pilot study the large surface area and high surface reactivity of iron nanoparticles were exploited to transform chlorinated hydrocarbons into less harmful end products in groundwater (Zhang, 2003).
- **Fuel cells:** Engineered surfaces are essential in fuel cells, where the external surface properties and the pore structure affect performance. The hydrogen used as the immediate fuel in fuel cells may be generated from hydrocarbons by catalytic reforming, usually in a reactor module associated directly with the fuel cell. The use of nano-engineered membranes to intensify catalytic processes could enable higher-efficiency, small-scale fuel cells. It has also been found that hydrogen condenses to high density in single-walled nanotubes (Dillon et al., 1997; Heben and Dillon, 2000), which stimulated a great deal of interest among the fuel-cell researchers in the possibility of carbon-nanotube hydrogen storage. However, one recent study has shown that this is still very challenging problem (Kajiura et al., 2003).
- **Displays:** The huge market for large area, high brightness, flat-panel displays, as used in television screens and computer monitors, is driving the development of some nanomaterials. Nanocrystalline zinc selenide, zinc sulphide, cadmium sulphide and lead telluride synthesized by sol-gel techniques are candidates for the next generation of light-emitting phosphors. Carbon nanotubes are being investigated for low voltage field-emission displays; their strength, sharpness, conductivity and inertness potentially make them very efficient and long-lasting emitters. Both Samsung and Motorola have reported significant progress on their flat-panel display projects (Choi et al., 2003; Motorola Labs, 2003). The carbon-nanotube-based flat-panel displays appear to be realizable in the not-too-distant future.
- **Batteries:** With the growth in portable electronic equipment (mobile phones, navigation devices, laptop computers, remote sensors, etc.), there is great demand for lightweight, high-energy density batteries. Nanocrystalline materials synthesized by sol-gel techniques are candidates for separator plates in batteries because of their foam-like structure, which can hold considerably

more energy than conventional batteries. Nickel metal hydride batteries, made of nanocrystalline nickel and metal hydrides, are envisioned to require less frequent recharging and to last longer because of their large grain boundary (surface) area.

- **Magnetic materials:** Magnets made of nanocrystalline yttrium-samarium-cobalt grains possess unusual magnetic properties due to their extremely large grain interface area. This could lead to applications in motors and analytical instruments, such as magnetic resonance imaging (MRI) and microsensors. Nanoscale-fabricated magnetic materials also have applications in data storage. In the future, the devices on computer chips that currently operate using flows of electrons could use the magnetic properties of these electrons – called spin – with numerous advantages. Recent advances in novel magnetic materials and their nanofabrication are encouraging in this respect.
- **Nanoelectronics:** The first room-temperature nanotube transistor was reported by Dekker and co-workers in *Nature* in 1998 (Tans et al., 1998). They demonstrated that a single-walled nanotube, draped across two pre-patterned gold electrodes, could be turned on and off by applying a potential to a gate electrode structure below the nanotube. These experiments suggested that nanotubes might someday be used as the building blocks or interconnections in nanoelectronics applications. However, nanotube-based electronic devices appear to be a very challenging task. Carbon nanotube electronics still faces a number of serious problems, such as the fabrication of large-scale arrays of devices and the controllable growth of the metallic and the semiconducting carbon nanotubes before becoming a serious technological alternative to silicon devices. However, considering the short time carbon nanotubes transistors have been investigated so far, the results already achieved are very impressive and fuel the hope of powerful carbon-nanotube-based nanoelectronics.
- **Medical implants:** Current medical implants, such as orthopaedic implants and heart valves, are made of titanium and stainless steel alloys, primarily because they are biocompatible. However, in some cases metal alloys may wear out within the lifetime of the patient. Nanocrystalline zirconium oxide (zirconia) is hard, wear-resistant, bio-corrosion resistant and biocompatible. It therefore presents an attractive alternative material for implants.

- **Water purification:** Nanoengineered membranes could potentially lead to more energy-efficient water purification processes. These applications would result in incremental improvements in technologies that are already available.

Are nanoparticles toxic?

As nanomaterials find more applications, especially in nanoparticle products for people's daily use, the opportunities for people to touch those tiny particles are increasing. Are they always friendly to humans? Two very recent experimental studies answered in the negative. The first group of researchers concluded from their experiments with nanoparticles that both single- and multi-walled nanotubes have activated human platelets and stimulated aggregation; the same materials stimulated the blockage of the carotid artery in their rat model (Radomski et al., 2005). The second experiment showed that exposing human skin cells to multi-walled carbon nanotubes and multi-walled carbon nano-onions (carbon atom clusters with the shape of multi-shell fullerenes and looking like a common onion) arrest cell cycles and increase cell death. The researchers looked at gene expression profiles and discovered that exposure to the nanomaterials disturbs a number of cellular pathways (Ding et al., 2005). Combined with an earlier report that 'Buckyballs cause brain damage in fish' (Holmes, 2004; Oberdörster, 2004), at least three alarms of the toxicity of nanoparticles were sounded. All these findings suggest that regulations for the safe management of nanoparticles are needed.

INSTRUMENTATION AND NANOMETROLOGY

The ability to measure and characterize materials (determine their size, shape and physical properties) at nanoscale is crucial for nanotechnology. It is directly related to whether nanomaterials and devices will be able to be produced to a high degree of accuracy and reliability, and whether the applications of nanotechnology are realizable. At the current stage, the overall progress in this area can be summarized as follows: with the invention of many powerful instruments, the goal of imaging and manipulating atoms to construct the patterns at nanoscale has been achieved. The measurements and characterization of nanostructures in surfaces are progressing steadily in the laboratory. However, to meet the stringent requirements of nanotechnology-based industry, serious challenges for nanoscale instrumentation and metrology remain.

Major Instruments used in nanometrology

Three types of instrument are used in nanotechnology with many variations: transmission electron microscopy, scanning probe microscopy and optical tweezers.

Electron beam techniques

Transmission electron microscopy (TEM) is used to investigate the internal structure of micro- and nanostructures. TEM can reveal the finest details of internal structure, in some cases individual atoms. TEM and high-resolution transmission electron microscopy (HRTEM) are among the most important tools used to image the internal structure of a sample. Furthermore, if the HRTEM is adequately equipped, chemical analysis can be performed. The scanning electron microscope (SEM) uses many of the basic technologies developed for the TEM to provide images of surface features of a sample. The best spatial resolution currently achieved is of the order of 1 nm.

Scanning probe techniques

Scanning probe microscopy (SPM) uses the interaction between a sharp tip and a surface to obtain an image. The sharp tip is held very close to the surface to be examined and is scanned back-and-forth. The scanning tunnelling microscope (STM) uses a sharp conducting tip held sufficiently close to a surface (typically about 0.5 nm) so that electrons 'tunnel' across the gap. The method provides surface structural and electronic information with atomic resolution. The invention of the STM led directly to the development of other 'scanning probe' microscopes, such as the atomic force microscope (AFM). The AFM uses a sharp tip on the end of a flexible beam or cantilever. Unlike the STM, where the sample has to be conductive, an AFM can image insulating materials simply because the signal corresponds to the force between the tip and the sample. True atomic resolution can be achieved in non-contact mode AFM. And, as mentioned in the previous chapter, SPM has the ability to manipulate atoms one by one.

Optical tweezers (single-beam gradient trap)

Optical tweezers use a single laser beam (focused by a high-quality microscope objective) to a spot on a specimen plane. The radiation pressure and gradient forces from the spot creates an 'optical trap' that is able to hold a particle at its centre. Small interatomic forces and displacements can then be measured. Samples that can be analysed range from single atoms and micrometre-sized spheres to strands of DNA and living cells. Optical tweezers are now a standard method of manipulation and measurement. Numerous traps can be used simultaneously with other optical techniques, such as laser scalpels, which can cut the particle being studied.

Using these measurement instruments along with other analytic techniques and nano-fabrication tools (such as Focus Ion Beam machines), the information on the topography and defect structure of a surface or an interface over distances close to the atomic scale can be obtained – measurements of single organic molecules and of structures, such as single-wall nanotubes, are already made. This is leading to greater understanding of the relationship between form and material properties, and enabling the control of processes at nanoscale and the design materials with special properties.

Challenges to nanometrology

Present nanoscale measurements have little metrological underpinning and few standards to ensure their reliability and repeatability. For example, there is a lack of applicable force or mass instrumentation with sensitivity adequate for engineering at the nanometre scale. This has become one of the major factors hindering the progress of nanoscale manufacturing. Hence, one of the greatest challenges facing nanometrology is to develop low-cost, reliable instrumentation and to set up internationally accepted standards for measurement of nanoscale phenomena and for the characterization and manipulation of nanostructures.

BIO-NANOTECHNOLOGY AND NANOMEDICINE

The most complex and highly functional nanoscale machines are the naturally occurring molecular assemblies that regulate and control biological systems. Proteins, for example, are molecular structures that possess highly specific functions and participate in virtually all biological sensory, metabolic, information and molecular transport processes. The volume of a single-molecule bio-nanodevice such as a protein is between one-millionth and one-billionth of the volume of an individual cell. So the biological world contains many of nanoscale devices and machines.

Bio-nanotechnology is concerned with molecular scale properties and applications of biological nanostructures. By using nanofabrication techniques and processes of molecular self-assembly, bio-nanotechnology allows the production of materials and devices such as tissue and cellular engineering scaffolds, molecular motors, and biomolecules for sensor, drug delivery and mechanical applications. Bio-nanotechnology can be used in medicine to provide a systematic, as well as a screening, approach to drug discovery, to enhance both diagnostic and therapeutic techniques, and to image at the cellular and sub-cellular levels at a much higher resolution than that of magnetic resonance imaging (MRI).

Nanoscience in this area

The primary aim of much current research is to obtain a detailed understanding of basic biochemical and biophysical mechanisms at the level of individual molecules. This knowledge will allow the design rules of naturally occurring molecular machines to be determined, which may lead to new technological applications. The tools developed in recent years (such as SPM) allow the direct observation of the behaviour of single molecules within biological systems. Examples range from the relatively large (45 nm) rotary molecular motors that power bacterial flagella 'propellers' to the tiny enzymes, such as ATP-synthase (9 nm), that catalyse energy conversion in biological processes. The intricate sequence of changes in molecular structure that forms the basis of such bio-molecular machines can now be measured directly by using AFM and optical tweezers. The recent development of high-speed AFM has enabled real-time molecular movement within a molecular motor to be observed directly. Future bio-nanotechnology and nanomedicine devices may exploit many classes of functional biological materials.

One group of proteins that is attracting attention is that of membrane proteins. These too are a class of protein-based machines that regulate many physiological processes. They include ion channels that enable rapid yet selective flux of ions across the cell membrane, hormone receptors that behave as molecular triggers, and photoreceptors that switch between different conformational states by the absorption of a single photon of light, the process that is the basis of vision and photosynthesis. That approximately one quarter of all genes code for membrane proteins is evidence of their immense biological importance. It is estimated that they will be the targets of up to 80 per cent of all new drugs. Single molecule techniques for both observation and manipulation are now being used routinely to study the selectivity and gating mechanisms of ion channels, and their response to drugs.

Current and future applications

Bio-mimetic structures may be devised that are based on naturally occurring machines: examples include catenanes and rotaxanes, compounds that behave as rotary or linear molecular motors, respectively. Medical applications are particularly promising. Areas such as disease diagnosis, drug delivery and molecular imaging are being intensively researched. Medical-related products containing nanoparticles are currently on the market in the US. Examples that exploit the known antimicrobial properties of silver include wound dressings containing nanocrystalline silver, which release ionic silver over a sustained period of time to provide an extensive antimicrobial spectrum of 150 different pathogens (Sample, 2001).

Array technologies

The enormously powerful array technologies, which use relatively large biological samples at the micrometre scale, are continuously being enhanced for sensitivity, size and data analysis. The original DNA chip approach, which holds an array of DNA molecules on an inert carrier, is now routinely used in gene and protein analysis. In nanotechnology the trend is to push towards higher resolution and smaller sample volume. Lab-on-a-chip technologies, which are used for sensing and supporting disease diagnosis, are also currently in the micrometre range, but progress in nanofluidic systems may lead to the production of integrated nanoscale systems. Their range of applications could include their use in improved devices for detection of biological and chemical agents in the field (Royal Society & Royal Academy of Engineering, 2004).

Electronics and information and communication technology (ICT)

One of the objectives of bio-nanotechnology research is to use the highly specialized functionality of proteins in devices such as molecular sensors. One of the greatest challenges is to understand the fundamental electronic properties of such molecules and the mechanisms by which electronic charge is transferred between them and metals, semiconductors, and novel nanoelectronic components, such as CNTs. Progress in this area could allow these 'smart' molecules to be integrated into devices and networks for specific ICT applications. The realization of a protein-based transistor is a major scientific challenge. DNA itself may turn out to be a useful electronic material, though the weight of experimental evidence so far indicates that it is not a good electrical conductor. However, used as a template, gold- or silver-'coated' DNA nanowires can be produced, and integrated circuits using DNA interconnects have already been realised which use the information coded in the DNA. More recently, a simple method to create robust DNA 'pyramids' that self-assemble in seconds has been invented. Each side of the tetrahedral pyramid is made up of a double helix of DNA. The pyramids can then be joined together to make larger three-dimensional nanostructures on which to build molecular electronic circuits (Goodman et al., 2005).

Thin films and crystals of the membrane protein bacteriorhodopsin have already been demonstrated to have potential photonics applications, such as optically addressable spatial light modulators, holographic memories and sensors. The photosynthetic reaction centre in this protein, which is only 5 nm in size, behaves as a nanometre diode and so it may be useful in single molecule optoelectronic devices. For example, its integration with electrically conducting CNTs and nanometre electrodes could lead to logic devices, transducers, photovoltaic cells, memories and sensors.

Self-assembly

The top-down approach to nanofabrication has the advantage of being able to produce pre-determined structures, but as we will see below, it is by no means an easy job. Much attention is now being focused on processes that involve some degree of molecular self-assembly, and in this respect biological materials have remarkable advantages over inorganic materials due to the diversity of self-assembled structures they can produce. Evolution in the natural world has produced an astonishing variety of biomolecular devices, and compared with conventional technologies, many natural molecular devices display enormous functionality. Among the most outstanding examples of synthetic structures now being fabricated are DNA-based geometrical structures (including artificial crystals) and functioning DNA-based nanomachines.

Drug delivery

There is enormous potential for nanotechnology to be applied to gene and drug delivery. The vehicle might be a functionalized nanoparticle capable of targeting specific diseased cells, which contains both therapeutic agents that are released into the cell and an on-board sensor that regulates the release. Different stages of this approach have already been demonstrated, but the combined targeting and controlled release have yet to be accomplished. In this event the way will be opened up for initial trials, and the eventual approval of such techniques will be fully regulated as for any new pharmaceuticals.

A related approach already in use is that of polymer-based drug therapies. They include polymeric drugs, polymer–drug conjugates, polymer–protein conjugates, polymeric micelles to which the drug is covalently bound, and multi-component complexes, which are being developed as non-viral vectors for gene therapy.

Many of these materials are now undergoing clinical trials for a variety of disease states. Gene therapy, where the DNA has been packaged into a nanoscale particle, holds much promise for the treatment of genetic defects, such as cystic fibrosis and immune system deficiencies. Alternative non-viral bio-nanotechnology approaches are being actively studied, though none has reached clinical trials. Advantages of these approaches include the versatility of synthetic chemistry (which allows tailoring of molecular weight), addition of biomimetic features to the man-made construct, and even the possibility of including bio-responsive elements.

Drug discovery

Nanotechnology techniques offer the possibility of studying drug–receptor interactions at the single molecule level, for example by using optical tweezers and AFM, so that a more direct approach to drug discovery becomes feasible. This approach might also enable, for example, the discovery of disease at the single-cell level, long before physical symptoms are manifested. This has been achieved by monitoring changes in atomic forces or ion conductance of a single receptor or ion channel when a drug molecule attaches. However, the industrial process will require the development of large arrays of such instruments working in parallel to create a high-throughput screening capability.

Medical imaging

Non-invasive imaging techniques have had a major impact in medicine over the past twenty-five years or so. The current drive in developing techniques such as functional MRI is to enhance spatial resolution and contrast agents. Nanotechnology already affords the possibility of intracellular imaging through attachment of quantum dots or synthetic chromophores to selected molecules, for example proteins, and thereby allows the direct investigation of intracellular biochemical processes.

Implants and prosthetics

As mentioned above, some materials, such as nanocrystalline ceramics, have certain properties (hardness, wear resistance, biocompatibility, etc.) that render them usable as implants in the long term. The development of nanoelectronic systems with high detector densities and data processing capability may allow the development of an artificial retina or cochlea. Important progress is already being made in this area, but many issues must be resolved before they can become viable treatments. Similarly, the introduction of nanoelectronics will allow biological neural processing to be investigated at much enhanced spatial resolution. Neurons of rodents have already been grown on nanofabricated surfaces to form elementary neural networks in which electrical signalling can be measured. By sending and receiving electrical impulses from the network, it may be possible to understand how neurons create memory by their responses to different patterns of stimuli.

It is hoped that this research will help some visually impaired people to regain their sight, or that muscle function will be restored to sufferers of Parkinson's disease. However, these developments raise potential ethical concerns about human enhancement and the convergence of technologies, in particular the question of whether the availability of body alterations that enhance human performance might diminish the role of disabled people in society, and whether progress in information processing and data storage technologies combined with developments in neurophysiology could lead to the possibility of non-therapeutic enhancement of human performance.

INDUSTRIAL APPLICATIONS

Current industrial applications of nanotechnology lie mainly in the characterization of materials, the production of chemicals and materials, and precision manufacturing. In general, these applications represent incremental rather than truly disruptive advances.

Characterization

The characterization of materials – the determination of their shape, size, distribution, mechanical and chemical properties – is an important part of the industrial process. It serves two broad purposes: as quality control and as part of the research and development of new processes, materials and products. As mentioned above, nanotechnology 'breakthroughs' have occurred regarding the tools used to observe and measure properties and processes at nanoscale. Sophisticated tools (STM, AFM and TEM) enable surface and interfacial characterization of materials at nanoscale, allowing individual atoms to be observed and analysed. This is leading to greater understanding of the relationship between form and material properties, and enabling the control of processes at nanoscale and the design materials with specific properties. However, the commercialization of such advanced functional materials requires that they can be made in a predictable, reliable way, and in sufficient quantities. Until that is achieved, production will be limited to academia and R&D departments in industry.

Fabrication techniques

Many techniques are capable of creating nanostructures, but the manufacturing approaches fall in two principal categories: bottom-up and top-down.

Bottom-up manufacturing

Bottom-up manufacturing involves the building of structures atom by atom or molecule by molecule. To do so, one of three methods is used: chemical synthesis, self-assembly or positional assembly. Among them, positional assembly is the only technique in which single atoms or molecules can be placed deliberately one by one.

1. Chemical synthesis
   Chemical synthesis is a method of producing raw materials, such as molecules or particles, which can then be used either directly in products in their bulk disordered form, or as the building blocks of more advanced, ordered materials. Metal oxides (titanium dioxide, zinc oxide, silicon dioxide, aluminium oxide, zirconia

and iron oxide) are currently the most commercially important nanoparticles. They are available as dry powders or liquid suspensions. Most genuinely nanoscale and nanostructured materials, however, are still at the laboratory scale of synthesis (kilograms/day scale or even less). Even so, the shape and size of the produced nanoparticles are quite diverse.

2. The role of self-assembly

Self-assembly is a bottom-up production technique in which atoms or molecules arrange themselves into ordered nanoscale structures by physical or chemical interactions between the units. Although self-assembly has occurred in nature from time immemorial, the use of self-assembly in industry is relatively new. There is economic and environmental interest in processes through which materials or product components essentially form themselves, creating less waste and using less energy. However, current understanding extends only to the creation of fairly rudimentary systems. One potential processing technique, known as directed self-assembly, involves the use of an external force or field (for example, electric or magnetic) to accelerate the often slow self-assembly process; this technique is attractive in an industrial context. The most useful building block, CNTs, can be grown by several techniques. However, because of a lack of understanding of the growth mechanism, the selective and uniform production of CNTs with specific dimensions and physical properties has not yet been achieved. And production capacity is very low. Current production capacity for CNTs is estimated to be around 100 tons per year (Royal Society & Royal Academy of Engineering, 2004). Most of the capacity is estimated to be multi-walled tubes, with single-wall tubes accounting for about nine tons of capacity.

3. Positional assembly

The final bottom-up technique is positional assembly, whereby atoms, molecules or clusters are deliberately manipulated and positioned one by one. Techniques such as SPM or optical tweezers are used for this. But it is extremely laborious, which indicates that currently this method is not suitable as an atomic-scale industrial process.

Top-down manufacturing

Top-down manufacturing involves starting with a larger piece of material and then removing material from it to make a nanostructure. Precision engineering and lithography are two key techniques for top-down manufacturing that have been developed for and refined by the semiconductor industry over the past thirty years. Top-down methods offer reliability and device complexity, but they are generally higher in energy usage and produce more waste than bottom-up methods. The production of computer chips, for example, is not yet possible through bottom-up methods.

1. Precision engineering
   In general, ultra-precision engineering and manufacturing underpin much of the microelectronics industry in everything, including the production of the wafers used as substrates for computer chips, the mechanical stages to position the wafers, and the precision optics used to print the patterns on the wafers. Through a combination of many advances (use of advanced materials for cutting tools; very stiff, precise machine tool structures; new linear and rotary bearing designs; and sensors for size control combined with numerical control and advanced servo-drive technologies), ultra-precision machine tools are now capable of high performance to handle the manufacture of nanoscale devices with simple shape surface, and at rather low output levels. Very precise process and temperature control is needed to achieve this level of performance (the latter being of the order of ±0.01°C).

2. Lithography
   Lithography involves the patterning of a surface through exposure to light, ions or electrons, and then subsequent etching and/or deposition of material onto that surface to produce the desired device. The ability to pattern features in the nanometre range is fundamental to the success of the IT industry. Electron- and ion-based methods are both capable of making sub-10 nm structures, but they are too slow to be used directly in production. Optical lithography is used for production of semiconductor devices. Although it does not have the resolution of the beam-based techniques, it does provide rapid throughput and cost-effective

manufacture. The requirement for shrinking device structures to nanoscale has placed enormous technical demands on the optical lithographic process.

The overall estimation of nanostructure fabrication

From the above overview one can infer that, due to its many obvious advantages over the top-down methods, such as material and energy saving, the bottom-up approach of nanostructure fabrication should be the favoured candidate for industry's development of nanotechnology applications. However, the current performance of this approach obviously shows that it has not yet matured to the stage of large-scale fabrication. As noted above, the products fabricated with the bottom-up approach are still confined to the laboratory scale with its capacity of production far below the requirement for industry application. The fact that a gram of single-walled carbon nanotube costs much more than a gram of gold underscores the situation. By the same token, though many delicate nanostructures – such as the bio-mimetic tiny machines and the newly invented nanocars (Shirai et al., 2005) (Figure 2.2) – assembled in the laboratory using a bottom-up approach have shown promise for applications, there is a long way to go before the industrial fabrication of these structures can be achieved.

In contrast to the bottom-up approach, the top-down approach is a much more mature technique for nanostructure fabrication; and it is the basis of the large-scale fabrication of computer chips that are already at nanoscale. But this technique is essentially the extension of the microtechnology developed in the semiconductor industry. By applying it at nanoscale, there are even more stringent requirements for its operation. It is expected that, as soon as the chip size shrinks further to 45 nm or less, the silicon chip will reach its physical tolerance limit, the top-down approach in this sector will give way to the bottom-up approach. That is why the new version of the Semiconductor Roadmap began in 2006 to include single-molecule switches fabricated with the bottom-up approach like the post-silicon era choice.

A notable trend in nanostructure fabrication is the convergence of both the bottom-up and the top-down approaches, which already started at the dawn of the twenty-first century. This convergence is expected to lead to exciting, new hybrid methods of manufacture and more efficient fabrication of nanostructures and nanodevices.

Figure 2.2. Nanocar made in laboratory[2]

Source: Shirai et al., 2005. Copyright © Rice University

In summary, industrial applications of the bottom-up approach are in their infancy while the top-down approach has already found its place in computer chip fabrication. In the long run, however, the top-down approach will be replaced by the bottom-up approach or hybrid methods. The manufacturing industry based on nanotechnology in the strict sense has not yet come of age.

CONCLUDING REMARKS

To conclude we offer the following brief observations.

1. Nanotechnology is an intrinsically interdisciplinary field of S&T: its continued development will enormously expand human understanding of nature and greatly promote the social progress of humanity.

   Development in nanotechnology has already significantly blurred the boundaries between different scientific disciplines

2 The nanocar consists of a chassis and axes made of well-defined organic groups with pivoting suspension and freely rotating axes. The wheels are buckyballs, spheres of pure carbon containing 60 atoms. The entire car measures just 3–4 nm across, making it slightly wider than a strand of DNA. The use of spherical wheels based on fullerene C_{60} and freely rotating axes based on alkynes permits directed nanoscale rolling of a molecular structure. Adapted from http://www.nanotech-now.com/news.cgi?story_id=12107.

and united scientists and engineers from different specialties in a common field. In this truly interdisciplinary field, which links physics, chemistry, biology, engineering and other areas, a great many innovative results have been obtained over the past decade and even more exciting ones are expected in the near future. All of these results are paving the way for a new technological revolution. However, the progress made thus far is still mainly confined to research and test phases. Nanotechnology is still very much at a primitive stage of development; a full-fledged manufacturing industry based on nanotechnology has yet to become a reality.

2. Developing countries are playing a more important role in the development of nanotechnology. This is a significant trend that should bring benefits to the majority of the Earth's population and should help to close the economic gap between developed and developing countries.

Up to the end of the twentieth century, most advanced technologies were developed and dominated by the developed countries, which widened existing gaps in knowledge, technology and economics between rich, developed countries and poor, developing countries. However, in the emerging field of nanotechnology research, scientists and engineers from developing countries have made remarkable contributions and shown themselves to be strongly competitive in their research. Several developing countries, including China (including Taiwan, Hong Kong and Macao), South Korea, Brazil, Singapore and India, have introduced nanotechnology research initiatives, that have made steady progress, and government funding for those initiatives continues to rise. Judging by recent statistics on their contributions to nanotechnology (based on numbers of publications, citing rates, patent award numbers and economic impact), these and other developing countries are quickly narrowing the gap in rank between themselves and the most developed countries, such as the US, Germany, Japan, the Netherlands, France, Canada and the UK, which have insofar led the way in nanotechnology research. Other developing countries, such as Argentina, Thailand and Iran, have also initiated nanotechnology research with government funding.

The narrowing gaps between rich and poor countries regarding nanotechnology research are one of the most significant and encouraging trends in scientific research.

3. To safeguard the continued progress of nanotechnology, both the study of regulations to ensure the safe usage of nanomaterials and the establishment of ethical guidelines for nanotechnology are urgently needed.

As mentioned above, results from several studies have already provided evidence that certain nanomaterials can be harmful to animals and humans (Radomski et al., 2005; Ding et al., 2005; Holmes, 2004; Oberdörster, 2004). Though the research in this area has yet to be carried out on a large scale, these initial findings clearly show that it is imperative that the regulation of the safe use of nanomaterials be seriously investigated. Likewise, since, as mentioned, some applications of nanotechnology in medical treatments may give rise to ethical problems, their possible consequences must also be carefully considered. In view of the catastrophic effects on researchers from nuclear technology and exposure to radiation during the early stages in the development of this technology, it is both timely and wise to consider the establishment of ethical guidelines for nanotechnology research now rather than when it is too late.

BIBLIOGRAPHY

Bachilo, S. M. et al. 2002. Structure-assigned optical spectra of single-walled carbon nanotubes. *Science*, Vol. 298, p. 2361.

Chen, J. et al. 2005. Bright infrared emission from electrically induced excitons in carbon nanotubes. *Science*, Vol. 310, p. 1171.

Choi, Y. S., Cho, Y. S., Kang, J. H., Kim, Y. J., Kim, I. H., Park, S. H., Lee, H. W., Hwang, S. Y., Lee, S. J., Lee, C. G., Oh, T. S., Choi, J. S., Kang, S. K. and Kim, J. M. 2003. A field-emission display with a self-focus cathode electrode. *Applied Physics Letters*, Vol. 82, 3565–67.

Dillon, A., Jones, K. M., Bekkedahl, T., Kiang, C., Bethune, D. and Heben, M. 1997. Storage of hydrogen in single-walled carbon nanotubes. *Nature*, Vol. 386, p. 377.

Ding, L., Stilwell, J., Zhang, T., Elboudwarej, O., Jiang, H., Selegue, J. P., Cooke, P. A., Gray, J. W. and Chen, F. F. 2005. Molecular characterization of the cytotoxic mechanism of multiwall carbon nanotubes and nano-onions on human skin fibroblast. *Nano letters*, Vol. 5, pp. 2448–64.

Dubrovinskaia, N., Dubrovinsky, L., Crichton, W., Langenhorst, F. and Richter, A. 2005. Aggregated diamond nanorods, the densest and least compressible form of carbon. *Applied Physics Letters*, Vol. 87, pp. 83–106.

Goodman, R. P., Schaap, I. A. T., Tardin, C. F., Erben, C. M., Berry, R., Schmidt, C. F. and Turberfield, A. J. 2005. Rapid chiral assembly of rigid DNA building blocks for molecular nanofabrication. *Science*, Vol. 310, pp. 1661–65.

Heben, M. J., Dillon, A. C., Cheng, H. M. and Dresselhaus, M. S. 2000. Room-temperature hydrogen storage in nanotubes. *Science*, Vol. 287, p. 591.

Holmes, B. 2004. Buckyballs cause brain damage in fish. 29 March. *NewScientist.com*. http://www.newscientist.com/article.ns?id=dn4825&print=true (Accessed 20 December 2006.)

Kajiura, H., Tsutsui, S., Kadono, K., Kakuta, M., Ata, M. and Murakami, Y. 2003. Hydrogen storage capacity of commercially available carbon materials at room temperature. *Applied Physics Letters*, Vol. 82, pp. 1105–7.

Ma, C. and Wang, Z. L. 2005. Road map for the controlled synthesis of CdSe nanowires, nanobelts, and nanosaws: A step towards nanomanufacturing. *Advanced Materials*, Vol. 17, pp. 1–6. http://www.nanoscience.gatech.edu/zlwang/paper/2005/05_AM_3.pdf (Accessed 20 December 2006.)

Ming, W., Wu, D., Bethem, R. van and With, G. de. 2005. Super-hydrophobic films from raspberry-like particles. *Nano Letters*, Vol. 5, pp. 22–98. http://nanotechweb.org/articles/news/4/10/16/1/raspberry (Accessed 20 December 2006.)

Misewich, J. A., Martel, R., Avouris, P., Tsang, J. C., Heinze, S. and Tersoff, J. 2003. Electrically induced optical emission from a carbon nanotube FET. *Science*, Vol. 300, pp. 783–6.

Motorola Labs. 2003. Press release: Motorola Labs announces significant progress in carbon nanotube technology. 1 July. http://www.motorola.com/mediacenter/news/

detail/0,1958,2981_2436_23,00.html (Accessed 20 December 2006.)

Oberdörster, E. 2004. Manufactured nanomaterials (Fullerenes, C60) induce oxidative stress in the brain of juvenile largemouth bass. *Environmental Health Perspectives*, Vol. 112, No. 10, 1058–62.

Radomski, A., Jurasz, P., Alonso-Escolano, D., Drews, M., Morandi, M., Malinski, T. and Radomski, M. W. 2005. Nanoparticle-induced platelet aggregation and vascular thrombosis. *British Journal of Pharmacology*, Vol. 146, pp. 882–93.

The Royal Society & The Royal Academy of Engineering. 2004. *Nanoscience and Nanotechnologies: Opportunities and Uncertainties*. 29 July. London, The Royal Society & The Royal Academy of Engineering. http://www.nanotec.org.uk/finalReport.htm (Accessed 22 December 2006.)

Sample, I. 2001. Small visions, grand design. *New Scientist*, no. 2311, 6 October, p. 30. http://www.newscientist.com/article.ns?id=mg17223114.200&print=true (Accessed 22 December 2006.)

Shirai, Y., Osgood, A. J., Zhao, Y., Kelly, K. F. and Tour, J. M. 2005. Directional control in thermally driven single-molecule nanocars. *Nano Letters*, Vol. 5, pp. 2330–34.

Tang, Z. K., Zhang, L., Wang, N., Zhang, X. X., Wen, G. H., Li, G. D., Wang, J. N., Chan, C. T. and Sheng, P. 2001. Superconductivity in 4 Angstrom single-walled carbon nanotubes. *Science*, Vol. 292, p. 2362–65.

Tans, S. J., Verschueren, R. M. and Dekker, C. 1998. Room-temperature transistor based on a single carbon nanotube. *Nature*, Vol. 393, pp. 49–52.

Wang, F., Dukovic, G., Brus, L. E. and Heinz, T. F. 2005. The optical resonances in carbon nanotubes arise from excitons. *Science*, Vol. 308, pp. 838–41.

Zhang, W. 2003. Nanoscale iron particles for environmental remediation: An overview. *Journal of Nanoparticle Research*, Vol. 5, pp. 323–32.

Chapter 3

IDENTIFYING ETHICAL ISSUES OF NANOTECHNOLOGIES

Joachim Schummer

The discussion of the ethical issues of nanotechnologies faces three difficulties: public hype, unclear definition and the early state of nanotechnologies.

First, nanotechnology is surrounded by a great deal of hype, hailed as the revolutionary technology of the twenty-first century with the enormous potential to radically change everything, from industrial production to the way we live and see ourselves as human beings, and the power to be disastrous to humanity. These utopian and dystopian visions, which all originate in science fiction, have been reinforced by various groups, including futurologists, software engineers, investment consultants, religious sects and governmental agencies. Such visions have created exaggerated public hopes and fears, as well as ethical concerns, with the consequence that nanotechnology is discussed mainly in terms of the societal and ethical implications of these visions (Schummer, 2004b; 2005). The proliferation of such 'envisioned' issues makes it particularly difficult to identify and articulate issues from an ethical point of view.

Second, the definition of nanotechnology is anything but clear. So many definitions abound that the meaning greatly differs from country to country, from discipline to discipline, and from public to public. In addition, with the launch of national nanotechnology programmes and their huge budgets, almost any science and engineering discipline can jump onto the bandwagon of nanotechnology and attach the 'nano' label to their research. The fuzziness of nanotechnology makes it almost impossible to identify ethical issues with the desired precision.

Third, most of what is now called nanotechnology is still in the early stages of research. Yet identifying ethical issues of the technologies in their final form would require foreseeing the outcome of the research and development processes, which even the researchers are unable to do.

Because of these three obstacles, it is advisable to approach nanotechnology from a critical and broader perspective. Before discussing ethical issues, it is important to understand the social context in which nanotechnology has emerged, including its various traditions, meanings, stakeholders, propagators and critics, as well as what researchers are actually doing in their laboratories. In addition to various governmental reports (e.g. Roco and Bainbridge, 2001; European Commission, 2004; Royal Society & Royal Academy of Engineering, 2004; Paschen et al., 2004), an international community of philosophers, ethicists, historians of science and social scientists has been discussing ethical issues of nanotechnology in view of the broader context (e.g. Fogelberg and Glimell, 2003; Baird et al., 2004; Hayles, 2004; Schummer and Baird, 2006; Nordmann et al., 2006). Apart from various other debates, scholars are in agreement that nanotechnology has emerged continuously over several decades rather than discontinuously, is diverse rather than monolithic, is much more normal than visionaries claim, and that the important ethical issues are concealed rather than clarified by the current nano-hype.

This chapter provides a survey of our current understanding are the most important ethical issues of nanotechnologies, while also placing emphasis on global equity issues and the potential impact of these technologies on developing countries (Schummer, 2006; 2007). The chapter will conclude with some recommendations on how governments should deal with these issues. But first, what is 'nanotechnology'.

THE DEFINITION OF NANOTECHNOLOGY

There are at least three different ways in which nanotechnology has been defined, and each one shapes perception of ethical issues in a radically different manner. Unfortunately, more sophisticated definitions have been neglected thus far (Schmid et al., 2003).

Nominal definition

The first definitional approach defines a term by providing necessary and sufficient conditions – what philosophers call a *nominal definition*. In this regard, the standard definition states that nanotechnology is the investigation and manipulation of material objects in the 1–100 nanometre range so as to explore novel properties and develop new devices and functionalities that essentially depend on the 1–100 nanometre range. Whether this is intended or not, the definition covers all the classical natural science and engineering disciplines that investigate and manipulate materials or material objects, such as chemistry, materials science, solid state physics, pharmacy, molecular biology and chemical, mechanical, electric and electronic engineering. This is because almost every material is structured in the 1–100 nanometre range such that its structure in this range determines its properties and, technologically speaking, its functionalities. Table 3.1 lists some commonly known substances that have crystallographic lengths in the nanometre scale, including elements such as sulphur and ordinary substances such as sugar (glucose) along with the much celebrated 'nanosubstance' buckminsterfullerene, or C_{60}. If one sticks to that definition, one will not perceive any new ethical issues, simply because there is nothing new about nanotechnology other than the name. And in fact researchers from most of the science and engineering

Table 3.1. Examples of commonly known substances with crystallographic lengths in the nanometre scale

Substance name	Empirical Formula	Biggest crystallographic unit cell length
Formic acid	CH_2O_2	1.02410 nm
Buckminsterfullerene	C_{60}	1.40410 nm
Glucose	$C_6H_{12}O_6$	1.48400 nm
Gypsum	H_4CaO_6S	1.52010 nm
Vitamin C	$C_6H_8O_6$	1.71000 nm
Alanine	$C_3H_8ClNO_2$	1.75900 nm
Sulfur	S_8	2.43360 nm
Vanillin	$C_8H_8O_3$	2.50990 nm
Cholesterol	$C_{27}H_{46}O_1$	3.42090 nm
Vitamin D_3	$C_{27}H_{44}O$	3.57160 nm
Pepsin	unspecified	29.01000 nm

disciplines are now re-labelling their research 'nano' – and rightly so according to this definition – because that helps them to raise funding.

Teleological or visionary definition

The second definitional approach, called *teleological definition*, defines nanotechnology by its future goals. On a very general level, these goals can be values such as health, wealth and security, or relative values such as smaller, faster, harder, cheaper – but this remains very unspecific. Since Eric Drexler first introduced the term in 1986, teleological definitions of nanotechnology have come in the form of visions of a future technology to be developed that will radically change everything, from industrial production to the physical, mental and social conditions of human life. According to this approach, current research belongs to nanotechnology if it helps to realize the envisioned nanotechnology that in turn will achieve the prospective goals. Numerous visions of this kind are in circulation, particularly in the US, and more recently in Europe. Besides Drexler and many other software engineers who dominate the popular book market on nanotechnology with their fantastic visions of nano-robots that can do anything (from DNA-repair to immortality and from self-replication to the total destruction of all intelligent life by 'grey goo'), there is a proliferating nanoscience fiction field that has essentially inspired them (Napier, 2004). In addition, US agencies have fashioned their own nanotechnology visions, which range from the Drexler-like 'shaping the world atom by atom' to transhumanist-like visions of a 'convergence of nanotechnology with biotechnology, information technology and cognitive science' for the 'enhancement of human performance'.

If one adopts these visionary definitions, ethical issues of nanotechnology immediately arise because they are part of the definitions. Such visions are meant to stir emotions, hopes and fears, rather than knowledge. They tell us what we should desire (what is good) and what we have to fear (what is bad). Thus, according to the visionary definitions, the ethical issues of nanotechnology have all been identified already by science fiction authors and futurologists, and so there is nothing left for ethicists to do. The problem, however, is that these visions are scientifically implausible or at least unfeasible in the foreseeable future. And because these visions are hardly related to actual

R&D activities in nanotechnology, they distract from the much needed analysis of the real ethical issues.

Real definition

The third definitional approach, also called *real definition*, refers to a list of particular research topics that usually appear under the umbrella of nanotechnology in governmental research programmes, in nanotechnology research centres, in nanotechnology journals and at nanotechnology conferences. Table 3.2 provides a list of the most frequently mentioned research fields (see also Chapter 2 above).

Table 3.2. Research fields that are usually related to nanotechnology

• scanning probe microscopy • nanoparticle research • nanostructured materials, polymers and composites • ultra-thin coating • heterogeneous catalysis • supramolecular chemistry • molecular electronics • molecular modelling • lithography in the production of ITs (integrated circuits) • semiconductor research and quantum dots	• quantum computing • MEMS (micro-electro-mechanical systems) • liquid crystals • small LEDs (light emitting diodes) • solar cells • hydrogen storage systems • biochemical sensors • targeted drug delivery • molecular biotechnology • genetic engineering • neurophysiology • tissue engineering

These research fields belong to a wide range of disciplines, including microscopy, materials science and engineering, surface science, organic chemistry, quantum chemistry, electrochemistry, electric engineering, solid state physics, mechanical and chemical engineering, biochemistry, molecular biology, and physiology. It is difficult to say whether they have anything in common other than that they are topical, since they represent the latest developments in most science and engineering disciplines. Moreover, contrary to frequent claims, there is no particular interdisciplinarity in these fields, whereas the list as a whole is, of course, very multidisciplinary (Schummer, 2004a). Therefore, it is more appropriate to speak of nanotechnologies (plural) than of nanotechnology (singular).

If one adopts this definition, as will be done in the following, the identification of ethical issues is a big challenge due to the diversity of topics. Moreover, the list of topics varies from country to country and has changed over time and thereby come to include new research fields. From an ethical perspective, it is difficult to identify any one possible issue that would apply equally to all the fields. On the other hand, the list includes many long-term research projects, such as semiconductor research, catalysis and genetic engineering, some of which have long been discussed from an ethical point of view. Since it makes little sense to repeat, for instance, the ethical discussion of genetic engineering now under the umbrella of nanotechnology, such discussion will be omitted in the following.

A TYPOLOGY OF ETHICAL ISSUES

Given the fuzziness and diversity of the concept of nanotechnology and the current hype, it seems appropriate to divide the discussion of ethical issues into specific and general issues. Specific ethical issues of technologies arise from particular research processes, the technological products and applications, and the manufacturing processes from laboratory to industrial scale. General issues arise from the way in which nanotechnology programmes as a whole are launched, controlled, and governed, as well as from how they are situated in the broader scientific and societal context. It is understood, however, that, while this chapter focuses on ethical issues, many nanotechnologies promise to have beneficial results for society and thus can contribute to the overall well-being of humanity if the ethical issues are properly considered.

Specific issues

Although the six groups discussed in the following do not exhaust the full range of ethical issues of the diverse field of nanotechnologies, they represent the most important ones in my view and bring some order to the matter according to the following rationale. First, since nanotechnologies aim at both improved materials and improved devices, it is reasonable to divide the discussion into health and environmental issues arising from new materials *and* issues of control arising from new devices. Second, because medical and military applications of nanotechnologies are the two main R&D foci of public concern, it is appropriate to discuss them separately. Third, the global dimension of nanotechnologies and their impact on economies, particularly those of developing countries, requires a discussion of global equity issues both regarding the material and the intellectual sides of nanotechnologies.

Health and environmental issues of new materials

At this point the most urgent ethical issues are the possible health and environmental risks of nanoparticles due to a substantial gap in all national and transnational regulations. It has been known for centuries that particles of exactly the same chemical composition have different properties depending on their size and shape in the nanometre range. This includes mechanical, optical, electromagnetic, thermodynamic, chemical, catalytic and biological properties, as well as of course the way these particles can migrate in the environment and through biological membranes. In addition, the phenomenon can in principle be explained, though hardly predicted in real cases, by quantum mechanics, for surface atoms and bulk atoms have different electronic structures, and the smaller the particle, the higher number of atoms are on the surface. However, in stark contrast to scientific knowledge, national regulations for chemicals, consumer products and work safety disregard the size- and shape-dependence of properties and focus solely on chemical composition. This means that a substance could, for instance, pass the required toxicity tests for new chemicals if the tests are performed on large particles, even if small particles of the same substance are toxic.

Of course, nanoparticles are not new. There are natural sources, such as volcanic ash and aerosols, as well as anthropogenic sources, such as combustion and abrasion, which have provided long-term exposure to certain nanoparticles. However, R&D activities in the fields of nanoparticles and nanostructured materials and composites are now systematically exploiting the size- and shape-dependence of properties with the goal of large-scale industrial production of improved materials. Much of the ubiquitous talk of 'the next industrial revolution' refers exactly to that. The results will not only be new materials but also entirely new exposure to nanoparticles, both regarding quantity and chemical composition, from production, distribution, consumption and abrasion processes. Both the new industrial opportunities and the new health and environmental risks go hand in hand, so that it is deeply irresponsible to celebrate the new opportunities and simultaneously disregard the risks.

Because some products based on nanoparticles or nanostructured composites are already on the market and many more are in the pipeline, there is an urgent need to define new standards for testing the safety of these products and their abrasion and to make these standards the basis for new regulations. In addition, there is an urgent need for the much-too-long-neglected research in nanoparticle toxicology, as well as in methods for making nanoparticles safe by surface treatment or encapsulation.

Control issues of new devices

From lithography to molecular electronics, many nanotechnologies aim to make electronic computing devices smaller and faster. In addition, devices for signal detection and emission, for solar energy collection and storage, and for mechanical, electrical and chemical operations, are being miniaturized at the micrometre level. All these technologies together provide a toolbox for various devices or systems of devices that can perform increasingly complex tasks with increasing autonomy, such as radio frequency identifiers (RFIDs) and systems for ubiquitous computing. Because many nanotechnologies are involved in extending the toolbox, they have been called 'enabling technologies' and only as such are they subject to ethical issues about devices. Whereas the universal 'nanorobot' is science fiction, and will certainly remain so forever, multi-task devices are being scaled down from the millimetre to the micrometre level with the help of various nanotechnologies.

Continuous technological processes, such as miniaturization, usually do not pose new ethical issues unless they transcend thresholds defined by human capacities. For instance, beyond the threshold of human sense perception, devices may cause changes, intrude on privacy or build up a surveillance system without being detectable. At a certain level of complexity, devices may perform quasi-autonomous decisions that we thus far confine to human beings with moral and legal responsibilities. Or devices may interact with each other on a systems level such that their collective behaviour becomes unpredictable and uncontrollable.

In all these cases, new ethical issues arise if the devices get out of control and harm human beings without there being anyone who can be held responsible. To cope with the emerging issue of responsibility, strict regulations are required that define the level of necessary human control and the scope of allowed tasks by devices and assign clear responsibilities to the producers and users of devices. In addition, the development of new devices needs to be accompanied by methods and instruments for detecting and disabling these devices.

Issues arising from military applications of nanotechnologies
Since a large part of governmental R&D budgets for nanotechnologies in at least some countries has been spent on military applications, it is appropriate to deal with military aspects separately. Typical public concerns are about the development of new nanotechnologies-based weapons, particularly biological and chemical weapons and the miniaturization and automatization of fighting and control systems, which would not only pose new threats and undermine international conventions but might also induce a new arms race. However, since most of the military research is actually classified, many of the concerns are based on speculation, for which again science fiction provides a rich source of inspiration. On the other hand, because it undermines democratic technology governance and public trust, an important ethical issue is the fact that under the label of 'nanotechnology' aspects of research are increasingly being classified and thus kept apart from public control and are thereby becoming subjects of public fears.

Moreover, military interests in R&D have shaped the goals of nanotechnologies from the very beginning, which impacts on human values in civil society. The targeted convergence of nanotechnologies with bio-, info- and cognitive sciences and technologies for the 'enhancement of human performance' (Roco and Bainbridge, 2002) is based on very specific ideas of what human improvement means. Enhanced physical strength, bullet-proof cloths, enhanced sensual capacities in the infrared or other ranges, enhanced mental capacities through brain-computer interfaces, and so on, all may well improve the military performance of soldiers. However, although these goals might appeal to some individuals, civil societies are built on different human values and different human qualities than those needed for military operations. Thus the intrusion of military values into civil society is a harmful distortion because it devalues moral and social values and the corresponding human capacities on which every civil society depends.

Finally, R&D projects of 'human enhancement' will likely be first tested on soldiers because they are the targets and have limited civil rights. Human experiments with brain-computer interfaces are particularly risky because they might cause long-term physiological and psychological harm that cannot be foreseen from animal experiments.

Issues arising from biomedical applications of nanotechnologies

If one ignores the science fiction stories about DNA-repair robots, immortality and 'superintelligence', the application of nanotechnologies in medicine follows rather conventional paths. From targeted drug delivery and biochemical sensors for diagnostics to genetic diagnostics and therapy, all has been ethically discussed for some time and is mostly regulated by national laws. On the other hand, the mere fact that such stories are propagated in the struggle for research funds and publicity is a serious ethical issue, because it irresponsibly preys on and toys with the hopes of patients who believe that their serious disease could be healed by some technological miracle.

There are two aspects of 'nanomedicine', however, that reinforce existing challenges to the medical system (other aspects will be discussed in subsequent chapters in this volume). The first is the development of black-boxed devices for self-diagnosis and automatic self-medication; for instance, biochemical sensors measure blood concentration data that is electronically processed to calculate the required medicinal doses for automatic injection. Of course, such an automatic medical system raises the aforementioned issue of responsibility concerning devices and their failures. Moreover, it challenges the medical systems insofar as it requires redefining the skills, tasks and responsibilities of medical doctors and nurses, who in this case are literally replaced by automata.

The second aspect is the reinforced pressure on the medical system to move from healing diseases towards enhancing the physical conditions of their patients beyond the level of health. Although this aspect is only loosely related to nanotechnologies, it is part of the political agenda in at least some countries that want nanotechnologies to converge with other technologies for the 'enhancement of human performance'. Apart from some peripheral fields, such as cosmetic surgery and the doping of athletes, 'enhancement' has never been the task of medical doctors and researchers, whereas the pharmaceutical industry has increasingly focused on so-called 'lifestyle drugs' over the past two decades. Indeed, the idea of 'enhancement' undermines almost all medical ethics deliberation, which has always been based on the first principle that no harm be done unless it yields health. What kinds of risks are acceptable for test persons during the clinical phase and for patients during the treatment if the outcome is not health but 'enhancement'? Moreover, since the resources for medical research and treatment are limited, the 'enhancement' business will absorb medical capacities that will then be missing for researching and healing serious diseases. Without effective countermeasures, the resulting imbalance will be particularly at the expense of the poor, both on the national and international level.

Issues arising from the material resources of nanotechnologies

Because nanotechnologies are associated with smallness, people tend to overlook the fact that the industrial scale production ultimately consumes thousands of metric tons of material resources per year. Two ethical aspects are associated with the consumption of such resources. First, the consumption of materials should follow the principle of sustainability so that future generations will not suffer from a lack of resources and an abundance of unusable waste. This means that nanotechnologies that are favourable are those that avoid using critical material resources or even replace technologies that use critical resources. In this regard, many nanotechnologies that consume rare elements, as in specific nanostructured ceramics and in opto-electronics, are less favourable. It also means that products from critical resources should be easily recyclable. Here the massive trend in materials engineering towards nanostructured composites poses a big challenge, because composites are particularly difficult to recycle.

The second ethical aspect of the consumption of material resources is particularly important for developing countries. It happens that most of the world's critical resources, particularly metals, are found in developing countries and that their economies essentially depend on mining and exporting these materials to industrialized countries. A long-term trend has been to find substitutes for expensive, natural or foreign material resources. For instance, synthetic dyes were substituted for natural dyes in the late nineteenth century; synthetic ammonia was substituted for natural nitre from Chile as fertilizer in the early twentieth century; and plastics have been substituted for wood and metals since the mid-twentieth century. All these substitution processes have had drastic effects on local and national economies.

Many nanotechnologies clearly follow this long-term trend. For instance, because of their extraordinary electric properties, carbon nanotubes are expected to substitute for high-conductive metals (copper, silver, gold, etc.) in electronics. Organic semiconductors are meant to substitute for semiconductor elements, such as gallium, germanium, indium, cadmium, selenium, arsenic and antimony. A few examples illustrate the economic dimensions these substitution processes can have (USGS, 2006; Schummer, 2007). The world market for tungsten, the element used mostly for ultra-hard materials (tungsten carbide and nitride), was $1.35 billion in 2005, with 90 percent of the mining production and world resources in China – much of the current materials research in ceramics is aimed at substituting exactly for that. Most of the catalysts used in oil refinement, chemical industry processes and automobile air pollution abatement are based on precious metals, such as rhenium ($47 million, mainly from Chile, Kazakhstan and Peru), palladium ($1.3 billion, mainly from South Africa and Russia), and platinum ($6.2 billion, mainly from South Africa); it is the express goal of nanotechnological catalysis research to substitute for or at least to reduce the required amounts of these metals. On the other hand, there are also nanotechnologies that create new demands for materials. For instance, in opto-electronics (light-emitting diodes, liquid crystal displays, solar cells, etc.), the use of indium-tin-oxide nanofilms has recently created a huge demand for the element indium, which is mainly mined in industrialized countries, such as Japan, Canada and Belgium, and more recently in China, resulting in sharply raising prices on the world market presently of about $370 million per year.

Due to the popularization of science fiction stories, the potentially drastic effects of nanotechnologies on the economies of developing countries, and thus on the increasing economic gap between poor and rich countries, have been entirely ignored. There is an urgent need for an assessment of the world economic impact of each nanotechnology if the affected developing countries are to prepare themselves in advance and respond with specific economic and R&D programmes. Although it is difficult to give specific advice before specific assessments have been carried out, two general recommendations can be made now to developing countries with material resources. First, they should focus on R&D that makes use of their domestic resources. Second, they should also research possible technological substitutes for current technologies that depend on their domestic resources in order to buffer the economic effects that such substitutes could have. For instance, it would be advisable for South Africa, which has 89 percent of the world's platinum resources, to research both new technologies that make use of platinum and new catalysts that could substitute for the current platinum catalysts in order to avoid economic disaster if others find such a substitute earlier.

Intellectual property rights issues of nanotechnologies

Although it is not specific to them, nanotechnologies are emerging at a time when intellectual property rights are changing in Western countries, and such changes are having negative side effects on developing countries (Sampat, 2003). Two trends are important in this regard. On the one hand, the criteria for what is patentable or not have become increasingly liberal, such that even basic engineering knowledge and database knowledge can now be patented. On the other hand, new regulations both in the US and in some European countries require that university employees report their inventions or their patentable knowledge to their university administration in so-called 'disclosure reports' prior to publication. The administrations then decide on whether patents are filed in order to earn revenues from licences. In US universities, the patenting rate moved from 30 percent to 50 percent in the late 1990s, which brought an increase of the overall licence incomes from $200 million in 1991 to $1.4 billion in 2004 (AUTM, 2006).

Due to both trends, knowledge produced in academic institutions, formerly published in scientific journals to become part of the public domain, is increasingly protected by patents and licensed on the market. While this has actually fostered the technology transfer from universities to local business and opened up new income sources for universities, the jungle of licences has made industrial development much more complicated and expensive, because every bit of basic engineering knowledge must now be purchased. Developing countries, which have thus far benefited from public knowledge and who are less able to pay licence fees, suffer most from this recent development. Hence, while developed countries' new intellectual property rights policies support local industries, they also dramatically increase the technology gap between developed and developing countries from both sides.

General issues

Ethics education for science students

In most Western countries, nanotechnologies research programmes have been financed by the reallocation of research funds from more basic to applied research, which incidentally follows a long-term trend. Since this shift in research funding has been accompanied by a shift in education, at least at the graduate level, towards applied science and engineering, it is increasingly necessary that ethics components be integrated in science education. Whereas ethics education is now mandatory for engineering students in many countries, this is not so for science education because scientists are still predominantly considered to perform ethically 'neutral' research. Since nanotechnologies stand for and induce this shift, ethics education must play a role at least in all the fields related to nanotechnologies.

Technology governance

The recent launch of national nanotechnology initiatives in many countries exemplifies a problematic trend in national and international science policy that undermines models of democratic and deliberative decision-making. Use of the buzzword 'nanotechnology' tends to disregard the diversity of the technologies covered by the term and the question of which nanotechnology should be supported and which not. Instead of evaluating the various pros and cons of each technology, governmental programmes have helped to generate hype by making unsubstantiated promises about 'the next industrial revolution'. Recurrent references to science fiction and futurology have given rise to exaggerated hopes and fears on the part of the public that undermine deliberative technology assessment. Instead of allowing citizen participation in science policy decision-making, many countries quickly jumped on the bandwagon, without much debate, to avoid being left behind.

Global equity

Over the past two centuries, technology has played a major role in the international economy, both by helping countries to grow economically and by reinforcing the economic divide between poor and rich countries. Since nanotechnologies are emerging at a time of increased economic globalization, and since enormous R&D efforts are being made in both developed and developing countries, it is imperative that each nanotechnology be assessed with regard to their impact on the economic status of developing countries. Such impacts can be beneficial or harmful for developing countries, both as consumers and producers of nanotechnological products and as providers of material resources and dumps. Two specific issues that are particularly relevant to developing countries have been discussed above, material resources and intellectual property rights, but many other issues need to be investigated (Schummer, 2007).

RECOMMENDATIONS

Ethical issues emerge if the development of new technologies or their prospective products conflicts with a society's ethical standards. While governments cannot control the ethical standards of their society, they are required to minimize the conflict. They can do so in four ways: (1) by enacting regulations to protect people from risks; (2) by supporting research to provide necessary knowledge for deliberative decision-making; (3) by educating the public on the various pros and cons of the technology in question to enable educated public technology assessment; and (4) by involving citizens in technology governance to ameliorate the conflicts. In conclusion, I would like to recommend that governments take measures regarding each of these four so as to deal with the ethical issues of nanotechnologies discussed above.

Regulatory needs

Regulation of nanoparticles

At this point the most pressing ethical issue is the failure to establish new toxicological standards and regulations for nanoparticles in order to reduce health and environmental risks. Governments are advised to follow the precautionary principle and enact regulations before industrial production and marketing of nanoparticle products take place so as to protect workers, consumers, the overall population and the natural environment. Failure to do so not only runs health and environmental risks but also the science policy risk that, after the first toxicological scandal, everything associated with 'nanotechnology' will meet with broad public hostility. In that case, the beneficial results of nanotechnologies would also find little public acceptance, which could render the huge amounts of governmental funding a huge waste of public money. Because the health and environmental risk issues are very pressing, individual governments should start working now on the national level of regulations, while also working on the much slower international level.

Regulation of devices

As devices become smaller, smarter and integrated into systems, they are increasingly able to perform autonomous tasks without being detectable or predictable. This undermines the moral and legal systems that are based on the assumption that only human beings make autonomous decisions for which they can be held responsible. To avoid the erosion of the concept of responsibility, and thus the basis of law and ethics, and to retain human control, regulations are required that clearly define the level of necessary human control and the scope of allowed tasks by devices and that assign clear responsibilities to the producers and users of those devices. In addition, the more powerful a device, the more dangerous is the criminal misuse of the device; consequently, in drafting regulations, it is necessary to consider whether and how the public availability of such devices is to be limited.

Regulation of 'enhancement'

Although societies differ in how much they allow their citizens to harm themselves, there are fundamental ethical limits based on general human rights that need to be regulated by law in all countries. First, nobody should become a test person for 'enhancement' experiments for any kind of reward or because of any kind of social or psychological pressure. Second, no 'enhancement' experiments should be performed on mentally retarded persons or on uneducated persons who are not fully aware of the risks. Third, no 'enhancement' experiments should be performed that bear any risks for persons other than the patient. This clause rules out, for instance, neurosurgery enhancement experiments, because other people would have to bear the risks of mental damage. In addition, if regulations allow for some other 'enhancement' treatment, it should be ensured that this does not absorb capacities from the healthcare system.

Regulation of intellectual property rights

For global equity reasons, it is desirable that the two aforementioned trends in intellectual property rights, which move knowledge from the public domain to the market, are reversed. To give developing countries a fair chance in the R&D of nanotechnologies, at least basic engineering knowledge that is funded by public money should become public domain knowledge again. The reversion would require both legal action on the national level of developed countries and international agreements. In addition, international knowledge platforms should be established that facilitate the knowledge transfer to developing countries with a focus on technologies that meet the particular needs of developing countries.

Research needs

Integrated research

Much of current nanotechnological R&D seems to repeat the failures of the twentieth century in that it is far too focused on narrow-minded technological goals without considering the broader perspective, including unintended side effects and technological and political measures to avoid them. Thus there is a strong need for integrated research that includes ethical and sociological research to understand the impact of both the intended and the unintended technological results on society, and that combines goal-orientated and harm-preventing technological research. For instance, materials research in nanoparticles must go hand in hand with research in nanoparticle toxicology and encapsulation. The development of new devices must be accompanied by the development of methods to detect, disable and even destroy these devices. The development of new materials should not only focus on technical performance but also consider, from the very beginning, sustainability issues, such as recycling and available resources, and how the increased materials demand would impact on the global economy.

Research focused on societal needs

The current nano-hype has created a situation in which everything that bears the label 'nano' is considered important and increasingly receives research funding. However, for ethical reasons it would be desirable that public money is spent on research that addresses the particular needs of society. For instance, it is questionable whether so much military research is really needed rather than civil research, and whether 'human enhancement' is more important than the medical treatment of serious diseases. Especially developing countries, with their small research budgets, should be advised to carefully scrutinize the variety of nanotechnologies in order to select and support R&D of those technologies that meet the specific public needs and strengths of the country (see also Chapter 6 in this volume). For instance, countries with freshwater problems might focus on new technologies for water treatment; countries with high solar energy input may invest in photovoltaic R&D rather than importing energy or the technology; and countries with domestic material resources should focus on technologies that both use and potentially substitute for these resources.

Educational and policy needs

Hype is the enemy of deliberative technology assessment and governance. In many developed countries, nano-hype has generated uncritical attitudes, blind support of any research that bears the nano label, and the public's exaggerated hopes and fears, which draw on science fiction rather than actual R&D projects. If developing countries copy the nano-hype, an additional danger is that 'nanotechnology' will become a symbol of modernism and thus that the assessment of nanotechnologies will turn into a symbolic debate on modernism versus traditionalism.

Therefore all countries are advised to take measures to avoid or to reduce nano-hype. Public education needs to address this issue by explaining the diversity of nanotechnologies and by pointing out the difference between actual R&D projects and science fiction. In addition, mandatory ethics components should be integrated in engineering and science education to provide students with ethical skills that allow them to analyse, assess and communicate the ethical and societal dimensions of technologies.

Because technologies increasingly shape society and determine the way we live, the entire process of projecting, supporting, guiding and regulating technologies – that is, technology governance – has become a critical part of politics. Democratic societies need to adjust to this development. Rather than letting experts or administrators make the crucial decisions, technology governance needs a stronger democratic basis, including citizen participation, from the earliest step on, in identifying societal needs and possible technological solutions. Democratizing technology governance is the best way to ensure that emerging technologies are developed in accordance with the ethical standards of a society.

BIBLIOGRAPHY

Association of University Technology Managers (AUTM). 2006. *AUTM Licensing Survey: FY 2004*. Northbrook, IL, AUTM.

Baird D., Nordmann, A. and Schummer, J. (eds.). 2004. *Discovering the Nanoscale*. Amsterdam, IOS Press.

Drexler, K. E. 1986. *Engines of Creation: The Coming Era of Nanotechnology*. New York, Anchor/Doubleday.

European Commission, Community Health and Consumer Protection. 2004. *Nanotechnologies: A Preliminary Risk Analysis*. Based on a workshop in Brussels, 1–2 March. http://ec.europa.eu/health/ph_risk/documents/ev_20040301_en.pdf (Accessed 20 December 2006.)

Fogelberg, H. and Glimell, H. 2003. *Bringing Visibility to the Invisible: Towards A Social Understanding of Nanotechnology*. Göteborg, Göteborg University Press. http://www.sts.gu.se/publications/STS_report_6.pdf (Accessed 2007)

Hayles, N. K. (ed.). 2004. *Nanoculture: Implications of the New Technoscience*. Bristol, Intellect Books.

Napier, A. S. 2004. *Nanotechnology in Science Fiction*. www.geocities.com/asnapier/nano/n-sf/books.html (Accessed 20 December 2006.)

Nordmann, A., Schummer, J. and Schwarz, A. (eds.). 2006. *Nanotechnologien im Kontext: Philosophische, ethische und gesellschaftliche Perspektiven*. Berlin, Akademische Verlagsgesellschaft.

Paschen, H., Coenen, C., Fleischer, T., Grünwald, R., Oertel, D. and Revermann, C. 2004. *Nanotechnologie: Forschung, Entwicklung, Anwendung*. Berlin, Springer.

Roco, M. C. and Bainbridge, W. S. (eds.). 2001. *Societal Implications of Nanoscience and Nanotechnology*. Proceedings of a workshop organized by the National Science Foundation, 28–29 September 2000. Dordrecht, Kluwer Academic Publishers.

——— (eds.). 2002. *Converging Technologies for Improving Human Performance: Nanotechnology, Biotechnology, Information Technology and the Cognitive Science*. Arlington, VA, National Science Foundation.

The Royal Society & The Royal Academy of Engineering. 2004. *Nanoscience and Nanotechnologies: Opportunities and Uncertainties*. 29 July. London, The Royal Society & The

Royal Academy of Engineering. http://www.nanotec.org.uk/finalReport.htm (Accessed 22 December 2006.)

Sampat, B. N. 2003. Center for Science, Policy, and Outcomes (CSPO) (ed.), *Knowledge Flows, Innovation, and Learning in Developing Countries* Temple, AZ, CSPO, pp. 39–81. http://www.cspo.org/home/cspoideas/know_flows/Rock-Vol1-2.PDF (Accessed 20 December 2006.)

Schmid, G., Decker, M., Ernst, H., Fuchs, H., Grünwald, W., Grünwald, A., Hofmann, H., Mayor, M., Rathgeber, W., Simon, U. and Wyrwa, D. 2003. *Small Dimensions and Material Properties: A Definition of Nanotechnology.* Bad Neuenahr, Europäische Akademie zur Erforschung von Folgen wissenschaftlich-technischer Entwicklungen.

Schummer, J. 2004a. Multidisciplinarity, interdisciplinarity, and patterns of research collaboration in nanoscience and nanotechnology. *Scientometrics*, Vol. 59, pp. 425–65.

———. 2004b. Societal and ethical implications of nanotechnology: Meanings, interest groups, and social dynamics. *Techne: Research in Philosophy and Technology*, Vol. 8, No. 2, pp. 56–87.

———. 2005. Reading nano: The public interest in nanotechnology as reflected in purchase patterns of books. *Public Understanding of Science*, Vol. 14, pp. 163–83.

———. 2006. Cultural diversity in nanotechnology ethics. *Interdisciplinary Science Reviews*, Vol. 31, pp. 217–30.

———. 2007. The impact of nanotechnologies on developing countries. F. Allhoff, P. Lin, J. Moor and J. Weckert (eds.), *Nanoethics: Examining the Societal Impact of Nanotechnology.* Hoboken, NJ, Wiley.

Schummer, J. and Baird, D. (eds.). 2006. *Nanotechnology Challenges: Implications for Philosophy, Ethics and Society.* Singapore, World Scientific Publishing.

US Geological Survey (USGS). 2006. *Mineral Commodity Summaries 2006.* Washington, DC, USGS. http://minerals.usgs.gov/minerals/pubs/mcs/2006/mcs2006.pdf (Accessed 20 December 2006.)

Chapter 4

ETHICAL ISSUES IN NANOMEDICINE

Bert Gordijn

The term 'nanotechnology' encompasses the study, creation and application of structures with *at least one* dimension of 100 nm or below. If an object does not have any dimensions within this range, it is excluded from the nanotechnological field. This definition appears to have been adopted by the majority of contemporary nanotechnologists. It is also used in conjunction with the US funding programme, the National Nanotechnology Initiative (NNI).

Thus far utopian dreams and apocalyptic nightmares have dominated the ethical debate on nanotechnology. These radically opposing evaluations harbour the risk of conflicts and unwanted backlashes. Furthermore, many of these drastic views are still based on narrow and obsolete visions of a nanotechnology dominated by self-replicating assemblers and nanomachines. Hence, the present state of the ethical debate on nanotechnology requires the development of more balanced and better-informed assessments.

To develop more discriminating ethical assessments of nanotechnology which are better informed by actual scientific developments, it is advisable to concentrate on a particular field of nanotechnological research. After all, there are diverse fields of nanotechnological research which do not necessarily demonstrate identical or even similar objectives and ethical aspects, for example: (1) materials and manufacturing, (2) nanoelectronics and computer technology, (3) medicine, (4) aeronautics and space exploration, (5) environment and energy, (6) biotechnology and agriculture and (7) security. The ethical assessments of developments in these fields are

NANOTECHNOLOGIES, ETHICS AND POLITICS – ISBN 978-92-3-104051-1 – © UNESCO 2007

likely to differ since the objectives of and ethical problems encountered by each will differ (Gordijn, 2005). This chapter focuses on emerging applications of nanotechnology in medicine, first describing the main developments therein and then analysing the ethical problems likely to be arise with advances in nanomedicine.

EMERGING APPLICATIONS OF NANOTECHNOLOGY IN MEDICINE

Although the term 'nanomedicine' suggests a specific field, nanomedicine cannot be appropriately regarded as a sub-discipline of medicine, such as clinical genetics, dermatology and cardiology. Rather, nanomedicine refers to a vast array of research endeavours relating to emerging applications of nanotechnology in medicine.

Many scientists at universities and in private companies are currently occupied with a great number of potential nanotechnology applications within medicine. This research is already affecting a variety of medical sub-disciplines.

The pace of developments in nanomedicine has been astonishing. Thus far the US has the lead. For example, the National Institutes of Health (NIH) endeavour to support the further development of nanomedicine by establishing 'Nanomedicine Development Centers' staffed by multidisciplinary scientific teams. These centres are to serve as the intellectual and technological centrepiece of the NIH Roadmap's Nanomedicine Initiative (NIH, 2006). In May 2004, NIH held the first public Nanomedicine Roadmap Initiative meeting to discuss the NIH vision for these centres. Subsequently, in March 2005, the first issue of *Nanomedicine: Nanotechnology, Biology, and Medicine* was published, which is the first international, peer-reviewed academic journal to focus exclusively on nanomedicine. It is the official journal of the American Academy of Nanomedicine (AAN) and ventures to provide the latest information both in nanomedical research and clinical applications. The AAN is the first professional academic society devoted to promoting research in nanomedicine. It was launched earlier in 2005 and has held its First Annual Meeting at the Johns Hopkins University in Baltimore, Maryland, on 15–16 August 2005.

The first explicit views of nanotechnological applications in medicine were marked by optimistic expectations about sophisticated, computer-driven medical nanomachines capable of performing all manner of highly complex medical tasks (Feynman, 1961; Drexler,

1986). However, the vast majority of current nanomedical research projects focus on a great variety of simpler nanostructures. For instance, Freitas identifies ninety-six subcategories of current nanomedical research projects. Among them are projects as diverse as studies of raw nanomaterials, nanostructured materials, control of surfaces, nanopores, DNA manipulation and sequencing, tools and diagnostics, intracellular devices, BioMEMS, nanotherapeutics and nanorobotics.

To give but an indication of the vastness of nanomedical research, we shall sketch in this section four tangible examples of topics that are addressed in current nanomedical research: (1) ways of improving diagnostic procedures; (2) methods that will permit the exact directing of certain therapeutic agents to a particular target, thus *drug targeting*; (3) methods to improve cancer therapy; (4) and manufacture of implants with greater biocompatibility and lifespan. In addition to these examples, we shall give a brief description of the great hopes expressed by some authors that (5) complex, computer-controlled medical nanomachines will one day be produced.

Diagnostics

A great many scientists are currently conducting research into various 'nanotechnological' diagnostic methods. One example of such research focused on the potential use of magnetic, nanometric particles to diagnose certain diseases. The aim is to use these particles as a substitute for the radioactive or fluorescing markers currently used to detect certain diseases. As with conventional methods, this new approach would also deploy monoclonal antibodies produced especially for diagnostic purposes and that stick to specific antigens. Depending on the antibody used, it is possible to detect the presence of corresponding pathogens (such as herpes, hepatitis B or HIV-AIDS viruses), or the presence of certain hormones, which would indicate pregnancy, diabetes, cardiac infarction or various tumours. To mark the antibody molecules, this diagnostic approach would bypass conventional radioactive or fluorescing substances, using magnetic nanoparticles instead. The marked antibodies would then be introduced to a tissue sample. To establish whether any antibodies had stuck to an antigen, thus indicating the presence of a pathogen, the tissue sample would then be exposed to a strong magnetic field, and the nanoparticles on the marked antibodies would become magnetized. If any antibodies had connected with a specific antigen, they would generate a measurable magnetic field, indicating the presence of a pathogen. If, by contrast, the antibodies in the tissue sample found nothing to which to attach themselves – signifying the absence of a pathogen – the magnetized antibodies would spin quickly around in circles, which would prevent the tissue sample from delivering a measurable magnetic field (Chemla et al., 2000; Alivisatos, 2001).

A second example of nanomedical research in diagnostics concerns the possibility of working in the future with 'quantum dots' (Alivisatos, 2001; Randal, 2001; West and Halas, 2000). Quantum dots are nanosized semi-conductor crystals. When stimulated by the light waves of a laser, they glow in different colours, the individual colour depending on a quantum dot's size. For example, a quantum dot measuring 2 nm shines green; a 5-nm dot shines red (Wolfe, 2002). Once they have been suitably cased – for example, by a thin layer of silicone (West and Halas, 2000) – these nanoparticles can be bound to various molecules and used as markers. When bound to antibodies, they can be used to detect pathogens.

A third example of nanomedical research in diagnostics is the use of 'gold colloid nanoparticles' (West and Halas, 2000). This new method is used to investigate the existence of a certain genetic sequence ('target sequence') within a solution; it exploits the fact that gold colloid changes in colour from red to purplish-blue when it accumulates (Alivisatos, 2001; West and Halas, 2000). This method may be sketched as follows: Two sets of gold particles are used. In the first set, several short strands of genetic material are bound to the individual gold particles, the sequence of these strands being complementary to the sequence of half of the target sequence. In the other set, the individual gold particles are also bound to several strands, this time complementary to the other half of the target sequence. If the target sequence really is contained within the solution, both gold particle sets combine with it. Because each of the gold particles bears several strands with a sequence complementary to the target sequence, a network results that holds the gold particles tightly together. The consequence of this accumulation is a recolouring of the previously red solution, proving the existence of the target sequence (Alivisatos, 2001; West and Halas, 2000; Taton et al., 2000).

Our last example of nanotechnological applications in medical diagnostics again involves cantilevers. Here, several cantilevers are arranged like a comb and combined to form a sensor. This sensor enables one to ascertain whether various molecular structures are present in a single sample. The principle is as follows: The surface of each 'tooth' of the comb-shaped sensor is coated with strands of genetic sequences, which differ for each 'tooth'. If the sample contains genetic material with a sequence complementary to one or several of the genetic sequences coating the cantilevers, then that material will bond to them. This causes the 'tooth' in question to bend by several perceptible nanometres. From the overall pattern of bending, one can deduce from the sensor the presence of certain molecules in the sample. One advantage of this method would be the sensor's potentially high sensitivity. Another would be the speed with which the existence of various genetic sequences could be established – nanomechanically, without marking procedures – within a single sample (Alivisatos, 2001).

Drug targeting

After diagnostics, a second focal point of nanomedicine is the development of procedures to precision-steer certain agents – such as insulin, vaccines, growth hormones or genetic material – towards their target. Such 'drug targeting' promises to be extremely beneficial because drug therapy is usually aimed at only one organ, a few cell types, or certain structures within one cell type – as in gene therapy. In addition to hitting their actual target, conventional types of medication, such as pills and injections, often flood other tissue types and cells with their active ingredients, which considerably increases the risk of side effects. An ideal system of treatment allows both the accurate measurement of various essential metabolic parameters in the defective tissue or defective cell, and then the release of the precise amount of medication required in the appropriate location. This would be possible, however, only if computer-controlled drug-targeting systems of nanometric proportions were already available.

Compared with the ideal procedures just sketched, the systems currently under investigation are still primitive. Efforts to date have produced the following results: certain nanoparticles and nanocapsules can be manipulated to develop a preference for a particular milieu. This enables them to be transported more directly to defective tissues or cells. Active agents can be packed in nanoparticles or nanocapsules, which are then attached to antibodies or other marker molecules. These nanosystems are capable of locating specified tissues or cells.

Nanoparticles suited to the transport of active agents include 'dendrimers', which are synthetic, spherical polymers characterized by branched surfaces and hollow interiors. Researchers are exploiting the structure of these nanoparticles by affixing marker molecules to their surface 'branches'; the active agents to be transported are placed within the hollow interiors (Alivisatos, 2001).

Research into the possibility of developing 'intelligent' transport capsules is also under way. Ideally, such capsules would be capable of transporting active agents to a specific location within the body and then releasing them there. One example of current research in this area involves 'gold nanoshells'. These spherical, nanometric systems have silicone or gold sulphide cores and are coated in gold. Depending on their dimensions, they can absorb different penetrating light wavelengths. The absolute size of a gold nanoshell, as well as the specific ratio between the diametre of its core and the thickness of its superficial layer of gold, determine the light wavelength it can absorb (Randal, 2001; West and Halas, 2000).

This feature could be exploited for drug targeting purposes as follows. By using a specific type of gold nanoshell, the absorption of certain light wavelengths is facilitated, and thus also the assumption of a certain temperature. First, a marker molecule – for example, an antibody or enzyme – is attached to the gold surface of a nanoshell to ensure that the drug targeting system reaches its target. Second, the nanoshell is connected to a heat-sensitive capsule containing the active agent to be transported. If this photothermal nanosystem were to be introduced into the body, it would not release its contents until the nanoshell were exposed to certain light wavelengths, that is, until enough heat were generated to cause the capsule to collapse (Alivisatos, 2001; West and Halas, 2000).

Cancer therapy

For some time now, research into new therapies for cancer has included various nanomedical approaches. A common denominator among these approaches is their exploitation of nanometric particles to combat the cancer. One example of this new research is nanometric radiotherapy, which was developed by scientists at the Memorial Sloan-Kettering Cancer Center in New York. Here certain radioactive agents – for example, individual actinium 225 atoms – are inserted into molecules constructed especially for this purpose. The molecules are then bound to suitable antibodies, enabling them to penetrate the cancer cells (Randal, 2001).

Other nanomedical research into cancer therapy aims to employ special gold nanoshells to destroy a tumour. Using light, these nanoparticles – equipped with marker molecules linking them to the cancerous cells – may be able to heat up enough to destroy the cancerous cells bound to them, while not affecting any of the surrounding, healthy tissue (Alivisatos, 2001).

Biocompatibility of implants

When an implant is integrated in the body, the precise characteristics of its surface play a crucial role. It is important, for example, that the surface will permit cell colonization. The molecular composition of an implant surface determines whether cells will be able to grow or will die. If researchers succeeded in appropriately modifying implant surfaces, cell colonization could be positively influenced. Nanomedical research is currently investigating just such possibilities (Alivisatos, 2001; Taton, 2001). One approach in this area is a precise roughening of the surface of polymer tubes to a nanometric depth. This would permit colonization with endogenous endothelial cells, since the latter would find sufficient points on that surface to which they could attach. If such a modification were to succeed at nanoscale, plastic bypasses, for example, could be rendered more compatible with the human body. Another approach to surface modification is research into the characteristics of bioactive ceramics used in bone implants. The crystalline microstructure of the ceramic surface influences bone formation. Appropriate nanometric structuring causes growth to penetrate the ceramic surface and thus create a permanent bond (Alivisatos, 2001). A final example here is research into the development of biocompatible 'interim layers'. Located between the surface of the implant and the endogenous cells surrounding it, they could promote cell colonization on the implant surface and thus integration of the foreign body. Overall, nanometric surface modification techniques could thus contribute one day to implants being more biocompatible and consequently having longer lifespans.

Medical nanomachines

Research into the possible development of nanosized, computer-controlled machines to perform all manner of medical interventions is probably the most futuristic example of nanomedical research presented here. Complete concepts for nanomachines capable of acting independently and equipped with molecular motors, as well as special vehicles and sensors to locate desired targets, are already available on paper. In practice, however, a nanometric construction of this kind has yet to be realized. Eric Drexler (1986) is convinced that the manufacturing of nanomachines will be possible in the future, and that their use will become part of medical routines. Biological nanomachines – such as ribosomes, viruses and bacteria – have, after all, existed in nature almost since life began. Nature is teeming with highly-structured, supramolecular, machine-like constructions.

These biological nanomachines constitute an important source of inspiration for the conception and potential production of artificial nanomachines. Natural molecular structures – such as DNA or enzymes – are even considered to be viable components for the construction of artificial nanomachines. Many authors estimate, however, that it will be between twenty and fifty years before the first nanomachines are ready for use in clinical practice. On the other hand, a sizeable group of authors has fundamental doubts about the possibility of realizing nanomachines in the future.

Assuming that the manufacture of nanomachines really were a viable option, what would they look like? The outer appearance of these future, more or less specialized, medical nanomachines would depend heavily on their area of use. According to Freitas (1998b), for example, the size of a machine intended to travel along human blood vessels would probably be 500–3,000 nm. The size of nanomachines moving not through the blood but through human tissue would probably be larger, 50–100 micrometres. And nanomachines travelling along the digestive tract or the bronchial tubes could be larger still (Freitas, 1998b). All sizes and shapes are conceivable, depending on the special task each machine is expected to perform.

Freitas has already addressed theoretically the design of selected nanomachines (Freitas, 1998a; 1999). On paper he has designed various nanomachines and described them down to the last detail. For example, he has described and drawn precise construction plans for 'respirocytes'. These are nanomachines of approximately 18 billion atoms, arranged to form spheres approximately one thousandth of a millimetre in diameter. These spheres would function as a type of gas container, pressurized and equipped with small pumps, and able to transport oxygen throughout the body. Sensors would be responsible for the gas exchange (Freitas,

1998a). The 'respirocytes' could assume the function of the red blood cells (erythrocytes) entirely.

A second type of nanomachine described and drawn by Freitas comprises mechanical phagocytes, which he calls 'microbivores'. He believes that these nanomachines could be used in patients to combat a whole number of pathogens. Once a microbivore were introduced into a patient's blood circulatory system, it could detect and destroy any microbiological pathogens located there (Freitas, 2001). Freitas has also designed artificial blood platelets, which he calls 'clottocytes'. In his opinion, they would coagulate the blood more efficiently than their natural equivalents (Freitas, 2000).

As already mentioned, it is not yet possible to produce medical nanomachines. Only by putting theory into practice will it be possible to determine whether the many and various nanomachine designs on paper really can be realized as envisaged. Designs for all manner of medical nanomachines can already be viewed on the Internet (see, e.g., Freitas, 2001–2004).

PROSPECTIVE ANALYSIS OF ETHICAL ISSUES IN NANOMEDICINE

Fantasies about potential, future applications of nanotechnology in medicine hardly provide a solid basis from which to evaluate the ethical issues accompanying any advance in this field of research. In addition, the applications currently under investigation have thus far attracted little if any ethical reflection – at least not beyond the usual ethical canon of responsible development and appropriate use regarding preclinical and clinical research. It will thus come as no surprise that a serious debate on such ethical issues has yet to take place. Nevertheless, it would be prudent to begin addressing these issues now, even while lacking a solid basis, in the sense of anticipatory or prospective ethics. The following analysis of the ethical issues likely to accompany a further development of nanomedicine is meant to be a first step in this direction.

Presuppositions of the analysis

Our review of the main ethical issues that may arise is divided into problems expected to be encountered in the short, the medium and the long term. Before beginning this anticipatory analysis of ethical issues, however, it is best to make explicit our basic assumptions about the future development of nanomedicine. In the short term, we assume that the bulk of research efforts in nanomedicine will focus on the diagnosis, prevention and therapy of disease. In the medium term, after initial solid results in that area have been achieved, the focus of nanomedicine will likely – slowly but surely – shift more towards the development of enhancement technologies. In the long term, medical enhancement technologies will be very powerful; our efforts to change ourselves by means of nanomedicine however and whenever they like could lead to a situation where it is no longer fitting to speak of 'human beings' at all.

The foregoing description of developments to date in the field of nanomedicine provides a solid enough basis to substantiate our assumption about the focus of nanomedicine in the short term. However, the link between nanomedicine and the further development of enhancement technologies requires further explanation. These technologies are already being developed in several present-day medical fields, including gene therapy, tissue engineering, bioelectronics, psychopharmacology, neuroscience, cosmetic dentistry, cosmetic surgery and endocrinology. Developments in all these fields might be further promoted by emerging nanotechnological applications. Let us focus on a few examples.

Tissue engineering is an interdisciplinary field concerned with the breeding of biological substitutes to restore, maintain or improve tissue and organ functions. One example of research in this field is the laboratory production of complex, three-dimensional tissue – literally a copy of human tissue design and function. The complexity of this process requires the deployment of different scientific disciplines. Principles and methods from both the life sciences and engineering sciences (hence the name 'tissue engineering') play major roles. Nanotechnology is expected to enable further and faster developments in tissue engineering. For example, it may enable the development of better scaffolds. In certain tissue engineering approaches, templates are needed to guide new growth from isolated cells, for the latter do not possess an intrinsic tissue structure. A histoconductive support is thus introduced into the artificially enlarged cell cluster to guide the regeneration of tissue structure in that cluster. The support consists of an extra cellular scaffold with a corresponding (two- or three-dimensional) structure. The materials used for this scaffold have to be highly porous, biocompatible and biodegradable, and can be either natural or synthetic. The cell culture is pipetted onto this scaffold, which creates and maintains enough space for the cells to form. Its structure both controls development and permits diffusion of nutrients. Nanotechnology may enable the development of better quality scaffolds and thus promote the advancement of tissue engineering. If successful, tissue engineering would be increasingly capable of restoring cell, tissue and organ functions. However, tissue engineering may also be used eventually to enhance these very same functions.

Bioelectronics investigates systems with a mixture of electronic and biological components. In medical applications, electronic devices are often wholly or partly 'built into' human beings. Research findings have already led to clinical applications, for example cochlea implants in the ear. Bioelectronic systems integrate progress in prosthetic technology with that in computer science. Not surprisingly, this field has miniaturization particularly in ICT to thank for its recent progress. Consequently, nanotechnology is expected to promote further developments in bioelectronics. Regarding the optimization of the human mind, memory performance could be enhanced beyond all boundaries, as well as the human capacity for processing information. On this topic, Freitas has said: 'Consider that a nanostructured data storage device measuring ~ 8,000 micron^3, a cubic volume about the size of a single human liver cell and smaller than a typical neuron, could store an amount of information equivalent to the entire Library of Congress. If implanted somewhere in the human brain, together with the appropriate interface mechanisms, such a device could allow extremely rapid access to this information' (Freitas, 1998b). The application of different bioelectronic systems to improve humankind would accordingly result in an ever more symbiotic connection between the human biological system on the one hand and the various technical systems at work on the other. This phenomenon is referred to as the 'cyborgization' of the human body.

Cosmetic surgery is concerned with surgical interventions to embellish physical appearance; it has already found widespread clinical application. Although invasive procedures to beautify our physical appearance are quite old, modern cosmetic surgery really started to develop in the twentieth century. As a result of the First and Second World Wars, there was a strong development in plastic surgery. Soon two different branches emerged: reconstructive plastic surgery and cosmetic surgery. The first focuses on patients with tumours, burns, mutilations, hereditary malformations and so on. The main objective of this kind of plastic surgery is the well-known *restitutio ad integrum*, reinstatement of human wholeness or intactness. On the other hand, cosmetic surgery focuses on healthy persons who for some reason are dissatisfied with their looks. The further development of nanotechnology may lead to a vast array of nanotechnological applications in cosmetic surgery which would enable new and better ways of enhancing the beauty of our physical appearance.

Through the enabling and catalyst function of nanotechnology, the aforementioned medical fields – as well as a variety of other medical subfields – could develop extremely quickly, thereby enormously increasing medicine's enhancement capabilities in a number of areas, including sensory, motorial and cognitive abilities, mood control and physical appearance. In the long term, this may lead to medical nanotechnology's ability to transform human beings to such an extent that it could even cease to be appropriate to regard the transformed being as human.

Short-term ethical problems

Risks of nanomedicine

The public debate on the risks of nanotechnology in general has been overshadowed by apocalyptic nightmares based on Eric Drexler's early vision (1986) of a nanotechnology in which universal self-replicating molecular assemblers play a pivotal role. Although this early conception of nanotechnology is now regarded as outdated – among others, by Drexler himself (Phoenix and Drexler, 2004) – it has substantially influenced the public debate (Gordijn, 2005). Still more influential in this connection has been Bill Joy, co-founder and scientific director of the private company Sun Microsystems. In his well-known article from the web journal *Wired*, 'Why the future doesn't need us', he issued general warnings about the various future risks of nanotechnology (Joy, 2000). He finds it especially troubling that research is being conducted into ways of producing assemblers. These assemblers would possess the hazardous ability to reproduce themselves – hazardous because an assembler could rapidly reproduce and then perhaps go out of control. In time, nanomachines capable of reproduction could escape, reproduce and violate the biosphere, leading to a catastrophe (the so-called 'grey goo' scenario). Michael Crichton described just such a catastrophe in his 2002 novel *Prey*, thereby immortalizing the picture of nano-dystopia for the public arena.

However, as mentioned above, Drexler's early view of the central importance of self-replicating molecular assemblers to nanotechnology has been subjected to severe criticism. First, critics have denied that it is even physically possible to create molecular assemblers (Ashley, 2001; Smalley, 2001). Second, even if it were possible, critics argue that it would be easier and more efficient to develop molecular manufacturing without these devices (Phoenix and Drexler, 2004). In view of such criticism, the risks of self-replicating assemblers running amok can largely be regarded as the stuff of fiction.

The real risks of nanotechnology in general and of nanomedicine in particular are probably to be found in areas that have yet to be identified. Currently, the hub of risk research in medical nanotechnology focuses on the health effects of nanoparticles – so far without any definitive results. Clearly this research is extremely important, but it has also proven to be quite complicated. After all, it is not possible simply to extrapolate from what we know about microparticles to the nanodomain. First, the laws of classical physics are not readily applicable to nanoscale objects. Thus there can be unexpected emerging phenomena. Second, nanoparticles are a great deal smaller than microparticles. As a consequence, they have access to almost every hidden niche of the body. While this property makes nanoparticles particularly interesting in connection with drug delivery systems, this may also pose new health risks. Third, the reactivity of nanoparticles is enhanced due to their larger relative surface. All in all, specific health effects of nanoparticles are not only material-dependent but may also be affected by particle size, surface chemistry, crystallinity, shape, charge, solubility, porosity, durability and so on.

To reduce the risks of the use of a specific category of nanoparticles, it will be necessary to conduct further research into its specific risk profile, that is, into the nature, magnitude and probability of the risks. Such research might result in adequate risk assessments, which could then serve as a basis for risk-reduction strategies. The development of feasible options for effectively dealing with risks might involve, for example, the substitution of a dangerous nanoparticle by a less dangerous one or a reduction of exposure to a dangerous particle.

When it comes to the ethical evaluation of the risks involved in applying nanotechnology in medicine, fortunately various well-established ethical evaluation criteria already exist. In medical research, for example, it is widely accepted that risks ought to be eliminated as much as possible in preclinical research by doing animal and other kinds of experiments. Furthermore, the chief norms with which nanomedical applications must comply are informed consent and proportionality between the risks and benefits.

Diagnosis without treatment

Diagnosis and therapy have a long tradition as medical goals. Ideally, the first serves the latter, which in turn serves to improve a patient's state of health or – should this prove impossible – to alleviate his or her suffering. Performed correctly, diagnosis therefore also indirectly contributes not just to therapy but also to a state of well-being. Like therapy, it therefore possesses an extrinsic value. Consequently, both of the cited objectives underlying nanomedicine are fundamentally desirable.

Another danger should not be ignored, however. As with genetics, the field of nanomedicine could also see an ever-widening gulf emerging between diagnostic and therapeutic capabilities. In clinical genetics, there is already a considerable gulf between diagnosis and therapy – for example, in connection with Huntington's disease or Alzheimer's disease. And it appears that in the future, as knowledge in genetics expands, more and more prognostic diagnoses will become possible for diseases for which there are no treatments yet available. This could increasingly lead to the situation where a patient is informed of an unfavourable genetic structure with a concomitant disposition to develop a certain disease and is then unable to receive treatment if the disease in question were to surface. The benefits of receiving this information can often be outweighed by the concomitant disadvantages.

Research in the area of nanomedicine could also develop in a similarly undesirable direction. The greatest efforts should therefore be made to prevent this from occurring. Diagnostic procedures should only be undertaken, for example, if they are clearly advantageous to patients. A detailed debate should be initiated to determine criteria to identify this clear advantage. Where possible, the development of new diagnostic procedures should be aimed at diseases that can already be prevented or cured. Since successful diagnosis of an incurable disease is often the first step towards developing a cure, however, this recommendation might prove somewhat problematic.

Medium-term ethical problems

Confusion regarding the concept of the body

For a long time the concept of the body used in medicine was thought unproblematic: 'the body' was everything inside the skin encasing an organism. The skin formed a clear boundary between the body and its surroundings; and everything outside this boundary unmistakably did not belong to the body. The concept of the body has recently become more complicated due to various medical achievements. The new discipline of organ transplantation, for example, has facilitated the implantation of 'subparts' in the body which were previously located outside the body. The same has happened with prostheses. Now it is unclear whether an implanted organ or prosthesis becomes part of the body upon implantation or remains foreign to the body. Equally unclear is the case of devices that are connected to the body externally in order to assume essential physical functions, for example a dialysis machine or a heart-lung machine. Here it becomes more difficult to say precisely what constitutes the 'body'.

It is to be expected that further developments in the field of tissue engineering will complicate the concept of body still more. To what category would an organ belong when it is grown from a cell sample taken from the recipient's own body (such that it could be termed 'endogenous') but developed outside the body and then implanted (such that it could just as easily be termed 'exogenous')? This would be further exacerbated if tissues and organs that had been developed in the laboratory were to cease, over time, to be distinguishable from their endogenous equivalents in appearance and function – as would in all likelihood occur. Under the influence of tissue engineering, the distinction between 'natural' bodies (without any interventions of the type described) and 'manipulated' bodies (containing artificial subparts) would become increasingly blurred. At some point that distinction may even become non-existent.

Developments in cosmetic surgery might add to the confusion. The subjective body image plays a pivotal role in cosmetic surgery. Although healthy, dissatisfaction with a certain aspect of one's physical appearance that is deemed to be suboptimal or even unmistakably ugly can seriously affect quality of life. However, the relation of the subjective body image to the objectively visible form of the body is not always straightforward. Subjective perception is influenced by all kinds of experiences, feelings of self-esteem, expectations in one's social environment and other socio-cultural factors. An extremely advanced cosmetic surgery might eventually transform the body into an object very similar to clothing that can come into and go out of fashion.

Furthermore, bioelectronic systems to improve normal sensory and motorial properties will probably become increasingly common. For example, there is speculation now about connecting up the natural eye to an artificial eye. Ideally, the artificial eye would have better and more universal vision than the natural one. With such an eye, human beings might be able to see wavelengths previously invisible to them (e.g. infrared) or things too far away or too microscopically small for the natural eye to see. As for the improvement of human motorial abilities, research is being done into 'supernaturally' fast running or 'supernatural' muscle power. With the help of such systems, soldiers could better defend themselves against attackers or flee faster. Widespread and invasive use of such bioelectronic systems that will enhance central bodily functions would almost certainly contribute to confusion over what parts or functions genuinely belong to our body. Thus it will become increasingly difficult to distinguish between the functions of one's own body and those of technology.

Changing attitudes towards the body

For some time now, new medical options – such as in transplantation medicine or cosmetic surgery – have been helping to mould new attitudes towards the body. Firstly, the various new therapeutic and cosmetic options available are increasingly giving the impression that the body is a whole composed of many different components. Physical development, physical functions and specific physical characteristics are beginning to appear less and less fixed by nature and unalterable, and more and more controllable and alterable. The view that is becoming dominant is that good functioning can be achieved by regularly replacing any defective parts. Increasingly, this changing attitude supports the view of the body as a product of technology, equivalent to a machine.

Secondly, a process of commercialization and commodification of the body can be observed. The body and all its subparts are increasingly becoming tradable goods. Cells, tissue and organs, as well as the 'borrowing' of a uterus, have all become tradable and purchasable. Not only egg cells and sperm, but also hair, blood, umbilical cords, placentas and foreskins are increasingly becoming part of an emerging economy surrounding the human body. Scientists and biotechnological companies are patenting genes. Embryos and brain-dead human beings are viewed as sources of replacement organs and tissue. The body is being treated as if it were the sum and source of parts, much like an inanimate machine. This process seems to be causing the human being to be stripped of the respect or even hallowedness it was formerly accorded.

Nanotechnologically enabled progress in medicine could accelerate this change in attitudes towards the body. More specifically, the commercialization and de-hallowing of the human body and all its parts could be intensified by the following developments in nanomedicine:

1. More and more, the body will become a product of technology. For example, complete warehouses of nanotechnologically manufactured biohybrid replacement cells, tissue and organs could be set up, from which people acquire the material they deem necessary.
2. The body will progressively become a part of technological systems. For instance, the body and its functions might constantly be checked and monitored with the help of nanosensors and nanochips, which would register all manner of emerging health threats. These gadgets might be linked with computer systems which would enable automatic responses for a broad range of common disorders and ailments (for instance, by activating certain nanometric drug release systems). Instead of feeling in charge of our own health, we might entrust technology more and more with this responsibility.
3. The body will increasingly be imitated by technology. With the rise of nanotechnology and the inherent tendency for miniaturization in a range of technical fields, there will be a tendency to copy the body's structures and mechanisms, such as the DNA, mitochondria or ribosomes in our cells or the neuronal networks of our brains.

Privacy and autonomy

A potential problem resulting from the nanotechnologically enabled cyborgization of the human body is its infringement on autonomy or privacy. Standard use of bioelectronic interventions to improve cognitive ability in human beings would intensify their links to computers. For example, contact to diverse databases would no longer be made via the fingers on a computer keyboard, but directly via the brain, which would be permanently connected to selected databases through an built-in bioelectronic information system. Standard use of enhancing bioelectronic interventions in human beings would also facilitate intensive links to other human beings. Communication with geographically remote people would then be possible not only via media such as the telephone or television but also permanently accessible – using the cyberthink communication system, for example – via direct brain-to-brain linkages.

Various side effects are conceivable in conjunction with these applications which would facilitate easy access to the privacy of other human beings. The digital fragments left over every time a computer is used could help to retrace and register the exact movements and actions of a human being. Furthermore, the extreme networking of human brains would enable their visual impressions to be registered, their thoughts to be recorded, etc. At any one time, a selected individual could thus be localized and registered in all respects – including his 'inner life'. Easy access to all areas of human privacy, even the most intimate, would then pave the way to (more or less) subtle influence and control. Direct networking of human brains would mean that human beings could receive – and subconsciously be influenced by – subliminal information. Yet this would infringe not only upon the person's privacy but also upon his autonomy.

Personal identity

When a human being thinks about his own person, he typically examines issues such as: What am I good at? How do I perform? Do I act responsibly? What is my particular character? What makes me unique? What can I remember? What is my life history? If a human being were constantly subjected to new bioelectronic interventions to improve his sensory, motorial and/or cognitive abilities, it could become increasingly difficult for him to answer such questions. Improving a human being's cognitive abilities, for example, could cause his own mental efforts to cease and thus to become totally dependent on high-performance bioelectronic implants. As a result, that human being would find it increasingly difficult to say which results of his thoughts and actions still constitute a *personal* achievement. If not only cognitive but also emotional improvement systems were to become available for implantation, it would become increasingly difficult to determine the characteristics specific to an individual. Moreover, if many different people were permanently connected to databases, the exclusiveness of possessing particular information would become relative, which in turn could reduce the uniqueness of each of those people thus connected.

Implantation of the cyberthink bioelectronic communication system – which would 'wire up' different individuals to enable them to exchange their conscious thoughts and experiences – could blur the line between the self and the cyberthink community. Once an individual as availed himself of such mental wiring, how will his own thoughts, experiences and life history to be kept separate from those of others? And the borders between the real and the virtual world would become increasingly blurred, which also would make it more and more difficult to determine one's own personal identity.

Medicalization

Widespread interventions to enhance sensory, motorial and/or cognitive ability, as well as our physical appearance, could be accompanied by the problem of medicalization. It is quite probable that once a certain number of people have undergone enhancing interventions, others would feel themselves under increasing pressure to do likewise. Without such interventions, they might fear not being able to keep up with those around them. In time, the attitude could prevail that, to be successful in life, one has to submit one's body to all manner of enhancing interventions. This would likely cause attitudes towards conventional human abilities to be quite negative. Average abilities could become almost akin to defects and requiring modification or even elimination. People could become afraid that their bodies and skills are fundamentally inadequate. As a result, enhancement interventions could trigger a process of medicalisation of what to this point have been regarding as ordinary functioning human abilities and normal physical appearances.

Social injustice

The further development and application of nanomedical enhancement systems could in time lead to social injustice. Clearly, modern society is very performance-oriented. Competition is so widespread in society that it is easy to imagine that individuals would be interested in nanomedical interventions which could improve their cognitive ability and thus give them a competitive edge, for such skills play a decisive role in maintaining and thriving in a professional career, amassing material wealth and gaining social prestige. It is also imaginable that parents would equip their children with nanomedical systems to enhance their cognitive ability and then continually update them to outdo the competition and ensure that their children attain long-term success.

Generally speaking, social injustice is the result of social differences that unfairly exist in a society. Nevertheless, social differences are not unfair *per se*. For example, it is often considered only fair that somebody who works harder and performs better should also earn more and be allowed to enjoy a higher standard of living. Unfair social differences only arise through an ethically unjustifiable distribution of social goods, such as status, prestige, money or employment.

Could the further development and application of nanomedical enhancement systems bring about such an unjust distribution of goods? To answer this question, two hypothetical scenarios may be distinguished: In the first, nanomedical enhancement interventions would be reserved for those who could afford to pay for them themselves; in the second scenario, these interventions would be made available to everybody, without exception.

In the first scenario, only the wealthy in a society would profit from the effects of enhancing nanomedical interventions, that is, only those who already enjoy a lofty social status and considerable material wealth. In this case, such interventions would cause the already unequal distribution of social goods to become even more imbalanced. Those already privileged within society would continue to reinforce and improve upon their already outstanding social and professional positions. Additional opportunities for social and professional improvement would be made available to precisely those people already at the top of the social hierarchy. This would be unjust because (1) it would further exacerbate the already unequal distribution of goods between the rich and the poor, and (2) it would obstruct within a given society the principles of equal opportunity and solidarity with the weak.

In the second scenario, where enhancement interventions would be made available to the population at large, there would be no exacerbation of existing inequalities regarding the distribution of social goods. Anybody interested in doing so could undergo an intervention and profit accordingly. This scenario would therefore not promote social injustice.

However, bearing in mind the present shortage of money in the health sector, the first scenario – in which nanomedical enhancement systems are reserved for the wealthy – would appear far more probable. Nobody is likely to want to cover the enormous costs of a population-scale improvement in cognitive abilities.

Changing view of the human being

Advances in nanomedicine could encourage a reductionist view of the human being as a complex molecular machine. This would mean that all physical and mental diseases and defects, as well as all desired human attributes, would increasingly be viewed under the paradigm of the molecular machine 'human being'. If this exclusively technical approach to such phenomena as disease, disability or self-perfection were to become dominant, any different though perhaps valuable – say, psychological or spiritual – approach to such phenomena could eventually be suppressed.

Ideas inspired by reductionism and technical optimism are to be found in the writings of Eric Drexler, one of the first authors to recognize and describe the countless potential applications of nanotechnology within the field of medicine (Drexler, 1986). In his view, the basic concept behind nanomedicine is extraordinarily simple: diseases, old age and other non-pathological human defects can all be viewed as unfavourable atomic and molecular configurations, which have different causes – for instance bacteria, viruses, genome mutations or accidents. Present-day medicine is also striving to eliminate unfavourable configurations, and yet its instruments to date are still comparatively primitive, including scalpels and drugs. If nanomachines were available, they could be used to eliminate these unfavourable configurations down to the last detail, as well as to restructure less favourable configurations as desired at a molecular level. The possibility of intervening directly at a such a level would considerably expand medical options for treatment. All manner of suboptimally functioning configurations could conceivably be optimized (Drexler, 1986).

If nanomedicine were to achieve its objectives, the medical field could be in danger of developing a dominant approach towards human beings, and thus its patients, that is one-sidedly reductive and technically optimistic. This view of human beings as molecular machines could come to dominate, as nanomedicine advances further, not only the realm of medicine but also the general public sphere. This may cause society, in its search for answers to dilemmas such as criminality or poverty, to fall prey to the temptation of narrowing its perspective to the purely technical. Such a development would be regrettable not least because it could easily suppress other valuable – such as political, sociological, spiritual or psychological – approaches to these problems.

However, if reductionism and technical optimism were indeed to gain a foothold in the medical field in the future, there would at least be hope that their inadequacies would eventually come to light. Extreme reductionism, for example, would soon reveal which diseases cannot be viewed and treated exclusively as the consequence of unfavourable atomic and molecular changes. Certain diseases, such as burn-out syndrome or anorexia nervosa, will probably never be fully reducible to the molecular level.

Long-term ethical problems

Blurred self-perception

If nanotechnology were actually to develop as successfully as many of its advocates hope, the symbiosis between man and technology would become closer and closer. The cyborgization of humankind would proceed rapidly. The process triggered by this cyborgization – namely human beings' becoming more and more a product of technology – can be referred to as human 'artefactualization'.

In recent times this process has been accompanied by a second process: technological systems are increasingly being developed along the lines of organic systems. This is happening in the fields of artificial intelligence, artificial life, robotics and neuronal networks, to name just a few examples. In the long term, the hope is that certain technological systems will be able to imitate various human or animal skills. Parallel to the artefactualization of human beings, this process can be termed an 'anthropomorphization' of technology.

The combination of these two processes – the artefactualization of human beings and the anthropomorphization of technology – could in the long term lead to the following problem: pairs of opposites which have existed for hundreds of years, including 'nature–culture', 'organic material–inorganic matter', 'conscious subject–unconscious object', could become fundamentally fuzzy. And yet such pairs had long represented essential elements of human self-perception. A fuzziness would engender a fundamental change in human self-perception to fit the new situation. Were the artefactualization of human beings and the anthropomorphization of technology to become a reality, what exactly would still belong intrinsically to the 'nature' of human beings, to their 'humanity'? What is it about human beings that may not be violated if we do not want to run the risk of having the 'typically human' disappear? What distinguishes human beings from artefacts? The very foundations of our view of humankind would be shaken under these circumstances.

Posthumanity

In the long term, nanomedicine could lead to a radical transformation of the human species. Humankind's striving to change itself however and whenever it wishes could ultimately lead to a situation where it is no longer fitting to speak of the 'human being' at all. Recent medical utopian fantasies describe how this process could unfold: with medical help, man departs from his human existence, with all its innate weaknesses and imperfections.

One example of such a fantastic surmounting of human being or nature would be the fusion of human being and machine. In another example, the human being is reshaped with the help of germ line genome modifications, performed consistently through successive generations. A third example is the so-called 'uploading' scenario, which would involve transferring the contents of the human brain to a computer. Using appropriately specialized nanomachines, the brain could be scanned at a sufficient resolution atom by atom. Then the neuronal networks of the brain could be implemented on an electronic medium and thereby bring into existence a software resident intelligence that could continue to exist *in saecula saeculorum*.

While these scenarios seem extremely speculative at present, it is nevertheless difficult to definitively refute the idea that nanomedicine might indeed one day come up with highly sophisticated and drastic enhancement technologies that could change humankind so radically that we could no longer speak of human beings in the conventional sense. Accordingly, it would be advisable to reflect prospectively on the implications of such a development. In this context, the following two questions should be addressed: (1) Which alterations would change human beings so radically that they could only be regarded as a post-human beings, and no longer as human beings? (2) Would it be ethically desirable for the human species to enter a post-human existence?

BIBLIOGRAPHY

Alivisatos, A. P. 2001. Less is more in medicine. *Scientific American*, Vol. 285, No. 3, pp. 59–65.

Ashley, S. 2001. Nanobot construction crews. *Scientific American*, Vol. 285, No. 3, pp. 76–77.

Chemla, Y. R., Grossman, H. L., Poon, Y., McDermott, R., Stevens, R., Alper, M. D. and Clarke, J. 2000. Ultrasensitive magnetic biosensor for homogeneous immunoassay. *Proceedings of the National Academy of Sciences USA*, Vol. 97, No. 26, pp. 14268–72.

Drexler, K. E. 1986. *Engines of Creation: The Coming Era of Nanotechnology*. New York, Anchor Press/Doubleday.

Feynman, R. 1961. There's plenty of room at the bottom. H. D. Gilbert (ed.), *Miniaturization*. New York, Reinhold, pp. 282–96.

Freitas, R. A. 1998a. Exploratory design in medical nanotechnology: A mechanical artificial red cell. *Artificial Cells, Blood Substitutes, and Immobilization Biotechnology*, Vol. 26, No. 4, pp. 411–30.

———. 1998b. *Nanomedicine FAQ*. http://www.foresight.org/Nanomedicine/index.html#NM FAQ. (Accessed 1 November 2005.)

———. 1999. *Nanomedicine* I: *Basic Capabilities*. Georgetown, TX, Landes Bioscience.

———. 2000. Clottocytes: Artificial mechanical platelets. *Foresight Update*, Vol. 41, http:// www.imm.org/Reports/Rep018.html (Accessed 1 November 2005.)

———. 2001. Microbivores: Artificial mechanical phagocytes using digest and discharge protocol. http://www.zyvex.com/Publications/articles/Microbivores.html (Accessed 1 November 2005.)

———. 2001–2004. *Nanomedicine Art Gallery*. Palo Alto, CA, Foresight Nanotech Institute. http://www.foresight.org/Nanomedicine/Gallery/index.html (Accessed 20 December 2006.)

———. 2005. What is nanomedicine? *Nanomedicine: Nanotechnology, Biology, and Medicine*, Vol. 1, pp. 2–9.

Gordijn, B. 2005. Nanoethics: From apocalyptic nightmares and utopian dreams towards a more balanced view. *Science and Engineering Ethics*, Vol. 11, No. 4, pp. 521–33.

Joy, B. 2000. Why the future doesn't need us. *Wired*, Vol. 8, No. 4. http://www.wired.com/wired/archive/8.04/ joy.html (Accessed 1 November 2005.)

NIH. 2006. *NIH Roadmap for Medical Research: Nanomedicine*. Betheda, MD., Office of Portfolio Analysis and Strategic Initiatives (OPASI), NIH. http://nihroadmap.nih.gov/nanomedicine/index.asp (Accessed 22 December 2006.)

Phoenix, C. and Drexler, E. 2004. Safe exponential manufacturing. *Nanotechnology*, Vol. 15, pp. 869–72.

Randal, J. 2001. Nanotechnology getting off the ground in cancer research. *Journal of the National Cancer Institute*, Vol. 93, No. 24, pp. 1836–38.

Smalley, R. E. 2001. Of chemistry, love and nanobots. *Scientific American*, Vol. 285, No. 3, pp. 68–69.

Taton, T. A. 2001. Nanotechnology: Boning up on biology. *Nature*, Vol. 412, 2 August, pp. 491–92.

Taton, T. A., Mirkin, C. A. and Letsinger, R. L. 2000. Scanometric DNA array detection with nanoparticle probes. *Science*, Vol. 289, pp. 1757–60.

West, J. L. and Halas, N. J. 2000. Applications of nanotechnology to biotechnology: Commentary. *Current Opinion in Biotechnology*, Vol. 11, No. 2, pp. 215–17.

Wolfe, J. 2002. A thousand dots of light. *Forbes/Wolfe Nanotech Report*, April. http://www.qdots. com/new/news/articles/0529. html (Accessed 1 November 2005.)

Chapter 5
ETHICS, NANOTECHNOLOGY AND HEALTH
Donald Evans

The emergence of new technologies which promise remarkable benefits for health has accelerated greatly over the last forty years. Perhaps the most notable among them have been the discovery of DNA followed by the mapping of the human genome and the invention of *in vitro* fertilization (IVF) techniques. These have, both separately and together, given rise to spin-off technologies and research endeavours that have stimulated much debate about ethical issues.

For example, in the case of genetics we have seen the possibility of matching pharmaceutical agents to patient groups with greater accuracy through the development of pharmacogenetics. This has raised important questions about the collection and use of human tissues and their exploitation in the production of drugs. Issues of privacy, consent and benefit-sharing have dominated these discussions. In the case of IVF, we have been forced to think carefully about the moral status of early embryos and whether they can be created for and used in research, bought and sold, selected for various qualities, and used for the benefit of others. The emergence of stem cell technologies has largely been stimulated by the availability of *in vitro* embryos. Ethical issues, such as protecting human dignity, causing harm and commodifying human beings, are still being debated on a very wide scale.

These two technologies have also joined hands in health intervention programmes, such as pre-implantation genetic testing and screening. Here questions about the notion of a disability and the dignity and value of the lives of people with disabilities have arisen, as have heated discussions about the differences between restoration of health and human enhancement. Are there important distinctions between making people better and making better people?

Into this already morally contested scene has come another technology promising much for the health of people, namely nanotechnology.

CLAIMS AND COUNTERCLAIMS

There are numerous difficulties attached to the ethical evaluation of this technology, for at this early stage it is not possible to distinguish easily between, on the one hand, the hype of the promises being made and the possibilities, both positive and negative, which are being discussed and, on the other hand, the realities of what is or what might become possible. Hence we have much to learn from nanotechnology's predecessors mentioned above.

Regarding genetics we might note that the new science held incredible promise for all sorts of stunning health interventions in the form of genetic engineering of cures for conditions such as cystic fibrosis. These might yet emerge, but developing them has turned out to be far more problematic than the early publicity suggested. While we have a remarkable array of diagnostic tools in the field, there is little, as yet, that can be done about righting the abnormalities deriving from the identified genetic faults, save by avoiding the birth of those who would be so afflicted. The latter intervention is itself a morally contested response to the health problems involved.

In the case of IVF, few people, if any, could have forecast the various developments that have occurred due to the availability of early human embryos in the laboratory. What was essentially a technique developed to relieve the distress of infertile couples has opened the door for vast extensions of knowledge in early human embryology and its various applications, such as the use of embryonic stem cells in research and, it is hoped, therapy.

Extreme optimism and extreme pessimism exists about the potential of nanotechnology. On the hyper-optimistic side, consider the following predictions:

> Nanomedicine will eliminate virtually all common diseases of the 20th century, virtually all medical pain and suffering, and allow the extension of human capabilities – most especially our mental abilities.

> Consider that a nanostructured data storage device . . . about the size of a human liver cell . . . could store an amount of information equivalent to the entire Library of Congress. If implanted somewhere in the human brain, together with the appropriate interface mechanisms, such a device could allow extremely rapid access to this information.
>
> But perhaps the most important long-term benefit to human society as a whole could be the dawning of a new era of peace. We could hope that people who are independently well-fed, well-clothed, well-educated, healthy and happy will have little motivation to wage war. (Freitas, 1998)

The most dire warning about the possibilities of the technology came from its founder, or at least from the scientist who invented the name for it (Drexler, 1986). He envisaged the creation of nano-robots that could be self-replicating and capable of consuming all organic life. While the author of this warning has lately somewhat distanced himself from that prediction (Phoenix and Drexler, 2004), his doomsday scenario, in which everything would become 'grey goo', has caught the imagination of numerous science fiction writers (Bear, 2005).

But there are more plausible and likely outcomes which we have to take into account when considering the relationship between nanotechnology and health. As a sobering corrective to the optimism regarding health benefits, consider the following possibilities envisaged by scientists working at Rice University's Center for Biological and Environmental Nanotechnology:

> One thing we've concluded is whatever these things (nanomaterials) are going to do, they're not inert. What will they do when they get into the environment, and what will they do when they get into people? (Vicki Colvin cited in Brown, 2002)
>
> Where does this stuff go? What will be its interaction with the environment? Is it the next best thing to sliced bread or the next to asbestos? We know nanomaterials have been taken up by cells. That sets off alarms. If bacteria can take them up, then we have an entry point for nanomaterials into the food chain. (Mark Wiesner cited in Brown, 2002)

A third scientist from this centre, Jennifer West, warns that proteins may attach to the surface of nanoparticles in the blood and thereby trigger dangerous consequences, such as blood clotting. Additionally, she claims that bacteria absorbing these particles might facilitate the entry of other dangerous materials into the body, and that toxins could similarly be spread in the environment (ETC Group, 2002). As we shall see, this area of risk becomes particularly significant when medical interventions might involve the introduction of nanoparticles into patients.

SETTING PRECEDENTS

There is more that we might learn from the ethical debate which has surrounded both IVF and DNA genetics. Many of the ethical problems thrown up by these technologies form the heart of ethical concern about nanotechnology in its health applications. In general it can be said that these are intensified in health nanotechnologies. Thus the debates in this field are not starting from square one as many of them did in IVF and genetics. When thinking about nanotechnology, we can learn from what has been said in those fields regarding both the identification of ethical issues and their treatment. Consider two examples.

First, we have already alluded to the development of pre-implantation genetic screening and testing, which has arisen out of the combination of the two aforementioned technologies preceding nanotechnology. It is promised that nanotechnology will provide us with methods of genetic examination which will be more rapid and vastly cheaper than those currently available. One of the practical limitations restricting the use of these technologies in population screening is its expense and the lack of sufficient expert resources to offer such a service. This barrier could well be removed in the near future by employing nanotechnology, thus increasing the pressure for such services to be provided. As a result, the major ethical problems of privacy and confidentiality, consent and ownership of genetic data, rights to information and dangers of genetic profiling, will be greatly intensified.

The second example is found in genetics where, as we have seen, much was promised in terms of engineering out genetic defects in developing human beings. Success in achieving such changes in the early stages of development would ensure better developmental health. But how much engineering is enough? Are we simply endeavouring to restore to an embryo or foetus its 'normal' potential? And how do genetic changes ensure the development of the same individual? Is it not a different person who results from such changes? Nanotechnologists have also forecast the remediation of health problems by nano-engineering. For example, it has been suggested, as noted above, that the insertion of intelligent nanoparticles into the brain might help restore a foetus with a developing intellectual impairment to its normal potential. But what is normality? Who is to decide this? If we could go beyond such gains and produce much greater potential for the developing child, should we do so? Questions about the proper role of medicine in this field have been discussed for some time in the context of genetic engineering. Is this eugenics? And if so is it wrong? There is little distinctive about the nanotechnological possibilities save for the scale of such extensions. Here we are once again called upon to consider the nature of persons and our moral reactions to engineered human beings, but possibly in view of more radical possibilities.

In considering the ethical dimensions of the development of nanotechnologies, we shall bear in mind the principles of the recently adopted UNESCO *Universal Declaration on Bioethics and Human Rights* (UNESCO, 2005).

NANOTECHNOLOGY AND PUBLIC HEALTH

Let us consider a few ways in which the developments in nanotechnology might impact on public health.

Nanotechnology as a threat to public health

Nanotechnology presents some ethical issues related to health which do not figure straightforwardly in either genetics or IVF-related technologies. These arise out of the introduction to the environment of nanosubstances that constitute environmental hazards. Nanotechnologies involve the production and use of new materials which, by their very nature, might present threats to the health of people. Insofar as this is the case, it has to be marked off from both the science of genetics and the creation of *in vitro* embryos. Here we find ourselves nearer to the development and uses of nuclear science, where the physical products produced and employed in the resultant technologies themselves constitute possible hazards to human health.

What are these products and how might they be harmful? Nanotechnology is concerned with incredibly small quantities of matter, as small as one billionth of a metre. Industry is interested in producing such particles of already existing compounds and such particles of modified materials that do not exist in nature. What is surprising to the ordinary person is that these very small quantities of materials behave quite differently from the larger masses with which we are familiar. We might expect them either to retain the qualities of the larger masses or possess them in diminished degrees. In fact, we now know that they might possess different qualities, such as colour, or the same qualities in greatly increased degrees, such as strength, chemical reactivity and electrical conductivity. Indeed, it is partly the latter fact which makes them so useful.

Whereas genetics has been said to deal with the building blocks of life, these nanoparticles have been said to be the building blocks of everything. Their minute character also enables them, when they enter the bloodstream, to cross the blood-brain barrier. They are so small that they can only be detected by the most sensitive of instruments and certainly not at all by ordinary people who might, in one way or another, be penetrated by them.

Their usefulness tempts us to employ them as soon as possible. For example, they are already used commercially in paints and cosmetics. They offer huge potential in being more effective catalysts than are currently used to assist in the refining of oil. Carbon nanotubes provide added strength and are planned to be used in the production of radial tires. Many companies are producing them by the ton. They are present, for example, in barrier creams to provide resistance to UV rays. Yet the US Food and Drug Administration (FDA) has no protocols for determining their safety, even though all products containing sunscreens are regulated as medicines in the US (ETC Group, 2002). That such products are already for sale does not provide any guarantees that they are safe. The FDA has approved zinc oxide for use in sunscreens without restriction on the size that can be used, despite evidence that it has phototoxic effects on cultured mammalian cells and their DNA *in vitro* (Royal Society & Royal Academy of Engineering, 2004, p. 44). The Scientific Committee on Cosmetic and Non-food Products (SCCNFP) has recommended that further research *in vivo* be conducted to clarify these findings and to produce reliable data on the absorption of zinc oxide through the skin (SCCNFP, 2003). European research has shown that there are grounds for concern in this case (Dunford et al., 1997). Until adequate risk assessment is performed, a question mark must remain over the release of these particles into the environment regarding possible threats they might pose to public health. Some warnings of this kind have appeared in the literature, for example:

> We consider that producers of nanomaterials have a duty to provide relevant toxicity test results for any new material, according to prevailing international guidelines on risk assessment. Even some 'old' chemical agents may need to be reassessed if their physical state is substantially different from that which existed when they were assessed initially. (Hoet et al., 2004)

Such materials can find ready access to the body. They also can find access into cells. From there on, we simply do not yet know what their effects might be. Environmental scientists are concerned because it is already known that some nanoparticles interact with biology in ways in which larger materials cannot (Brown, 2002). It has been proven that nanoparticles inhaled from the environment can be translocated to the brains of rats and that this probably occurs via the olfactory nerve (Oberdörster et al., 2004). It has been further shown that uncoated buckyballs (nanoparticles of carbon) cause damage to the brains of aquatic species by causing lipid and protein damage (Oberdörster, 2004). This kind of tissue damage has been linked to Alzheimer's disease.

Scientists are endeavouring to make such particles safe in some areas of application by coating them. This will be necessary if they are to be of use in the therapeutic field, which we shall discuss later. Policies for providing adequate reassurances concerning environmental risk must be produced before countries rush headlong into indiscriminate production and employment of nanomaterials. In this process of policy development, due regard must be paid to the following articles from the *Universal Declaration on Bioethics and Human Rights*:

> **Article 16** The impact of life sciences on future generations, including on their genetic constitution, should be given due regard.

> **Article 17** Due regard is to be given to the interconnection between human beings and other forms of life, to the importance of appropriate access and utilization of biological and genetic resources, to respect for traditional knowledge and to the role of human beings in the protection of the environment, the biosphere and biodiversity.

Nanotechnology as an aid to public health

There are three areas of public health which may benefit greatly from developments in nanotechnology. The first marks a direct counterbalance to the environmental threats noted above.

Nanotechnology as an aid to environmental safety

There are already many pollutants in the environment. Despite the concerns noted above about the possibility of pollution problems being worsened by the development and widespread use of nanoparticles, there is a real possibility that the employment of nanotechnology could remove pollutants from the environment more effectively than has been dreamt of previously.

The most outstanding opportunity with respect to public health probably lies in the provision of clean water. While the developed world might take the provision of potable water for granted, the number of deaths worldwide resulting from the consumption and use of impure water dwarfs that of any other single cause, including AIDS and famine. The World Bank estimates that over half the world's population lacks basic sanitation and some 1.5 billion people lack access to clean water. The result is that 80 percent of the diseases in the developing world are water-related. It has been estimated that these lead to 3.4 million deaths every year, the majority of victims being children (Foresight Nanotech Institute, 2006).

The contaminants take a variety of forms, both organic and inorganic. Bacteria and viruses play a major role; oil, other organic pollutants and heavy metals also play their part (Singer et al., 2005, pp. 58–59). Various promising nanotechnological solutions have been devised to address this problem, some of which are the following:

1. 'Intelligent' membranes can be produced to make affordable and portable filter systems which will remove most contaminants, including bacteria and viruses. These materials are 10,000 times more capable of binding bacteria and toxins than activate carbon.
2. Nanomagnets, with various coatings, can be designed to deal with specific contaminants, including oil, from water. Such dust-like preparations could be spread over wide areas and gathered up affording almost 100 percent effectiveness. These nanomachines could be recycled and used over and over.
3. Magnetite nanoparticles combined with citric acid could remove heavy metals from water.

The prospect of removing terrible diseases, such as cholera and typhus, as well as common gastroenteritis, from the lives of vast populations of sufferers is cause for great excitement. But will it happen?

Before that is possible, the developed world will need to pay considerable attention to the following articles from the *Universal Declaration on Bioethics and Human Rights*:

> **Article 10** The fundamental equality of all human beings in dignity and rights is to be respected so that they are treated justly and equitably.

> **Article 13** Solidarity among human beings and international cooperation towards that end are to be encouraged.

> **Article 14b** 2. Taking into account that the enjoyment of the highest attainable standard of health is one of the fundamental rights of every human being without distinction of race, religion, political belief, economic or social condition, progress in science and technology should advance... b) access to adequate nutrition and water . . .

> **Article 15** 1. Benefits resulting from any scientific research and its applications should be shared with society as a whole and within the international community, in particular with developing countries . . .

Yet it is likely that the financial rewards available to the nanotechnology developers will tempt them to concentrate on advances that will show the greatest economic returns for their investors. If this occurs, then the gap between the rich and poor nations will grow, and a remarkable opportunity to contribute to a better world will be missed. That choice has been starkly presented to the world by Singer et al. (2005, p. 64) as follows:

> Will nanotechnology produce the nanodivide? Resources might be directed primarily to nanosunscreens, nanotrousers, and space elevators to benefit the 600 million people in rich countries, but that path is not predetermined. Nanotechnology could soon be applied to address the critical health, food, water, and energy needs of the 5 billion people in the developing world.

Nanotechnology as an aid to epidemiology and public health provision

Epidemiology focuses on populations rather than individuals. It is concerned to make discoveries about the aetiology of diseases by studying trends, which, in turn, involves having ready access to health data on very large numbers of patients. As such it is has a public health interest, though its discoveries, of course, benefit individual patients in the long term.

Similarly, public health initiatives focus on what has often been referred to as the herd, as in 'herd immunity', rather than individual patients. An example would be the provision of immunisation or vaccination programmes, such as for protection against whooping cough. However, whenever a doctor faces the mother of a child to offer such protection, he or she cannot guarantee that it is in the individual child's best interest to be vaccinated since, sadly, on rare occasions adverse reactions to the vaccine can cause severe brain damage. Nevertheless, once a certain proportion of the population has been vaccinated, a herd immunity is produced and the possibility of an epidemic is eradicated, which is in the interests of all the children in a given community.

Likewise, when population screening programmes are conducted, such as in cervical cancer screening, it is important to be able to follow the unfolding medical history of those involved in the programme in order to evaluate its effectiveness. Once again, the focus of the exercise is not the welfare of individual patients but the overall success of the population-based programme.

How can nanotechnology offer support to such enterprises? There are various possible developments of nanotechnology in the field of diagnostics and information technology which, together, would offer great assistance to practitioners involved in the aforementioned kinds of activity. For example, the production and insertion into patients of nanocomputer-like particles is constantly canvassed as providing early warnings of the developments of disease conditions. This is not really surprising since proteins might themselves be regarded as nanodevices which perform highly specific functions in virtually all biological sensory, metabolic, information and molecular transport processes (Royal Society & Royal Academy of Engineering, 2004, p. 20). Nanotechnology simply seeks to replicate such devices at the molecular level. In addition, these nanocomputer-like machines can be subject to external control. For example, treating doctors can manipulate them in various ways to administer treatments to patients such as dispensing therapeutic substances as needed. This opens up the possibility of tracking them with remote sensing devices (Mertz, 2001). Such a possibility would provide major assistance in the audit of public health initiatives and in epidemiological research. However, given the size of the particles, people might not even be aware that they were carrying them about in their person or that they were under surveillance, perhaps regarding some of the most intimate and significant features of their lives. Such a scenario naturally raises important questions about privacy, confidentiality and consent.

The tension between the interests and rights of the individual patient, on the one hand, and the interests of public health, on the other, is not new. Discussion is called for to determine how the tension can best be managed so as to protect individual autonomy without rendering the public health initiatives impossible. Issues trented thereby include continuing consent, methods of approach to patients in gathering data, the usefulness of the distinction between the use of identifiable and non-identifiable data, restrictions of access to irrelevant information or information not canvassed in the original consent of the patient, and restrictions of access to unauthorised persons. The new data gathering powers offered by nanotechnology intensify these ethical discussions, for they are seen to pose greater threats to civil liberties.

In this respect, Articles 3 and 6 of the *Universal Declaration on Bioethics and Human Rights* will need to be taken into account in the formation of public policy on nanotechnology:

Article 3

1. Human dignity, human rights and fundamental freedoms are to be fully respected.
2. The interests and welfare of the individual should have priority over the sole interest of science and society.

Article 6

1. Any preventive, diagnostic and therapeutic medical intervention is only to be carried out with the prior, free and informed consent of the person concerned, based on adequate information. The consent should, where appropriate, be express and may be withdrawn by the person at any time and for any reason without disadvantage or prejudice.
2. Scientific research should only be carried out with the prior, free, express and informed consent of the person concerned. The information should be adequate, provided in a comprehensible form and should include modalities for withdrawal of consent. Consent may be withdrawn by the person concerned at any time and for any reason without any disadvantage or prejudice.

Nanotechnology as an aid to population screening

There is a further form of population surveillance that deserves separate consideration, namely community or population genetic screening.

This possibility, especially at the level of neonatal screening, has been canvassed with increasing regularity over the past ten years or so for various reasons (HGC, 2005). On the one hand, it has been claimed that wider coverage of such screening would empower people to make sensible lifestyle, social and domestic choices, including reproductive choices. On the other hand, it has been suggested that, as more information of this kind is placed in the hands of healthcare providers, a number of objectives would be realized, including a reduction in the number of births of children with major genetic disorders, the facilitation of better planning of efficient and effective healthcare delivery for the next generation, and assistance in the planning of research programmes.

Laudable as such objectives might appear, each of them poses considerable ethical challenges. However, above and beyond these, such programmes have been subject to the law of diminishing returns. For example, at present various genetic tests are provided to couples who are deemed to be at risk of having children with genetic disorders. These tests are expensive and highly specific. Expanding such provision to a wider range of people would be likely to return proportionately smaller numbers of positives. The wider the net is cast, the smaller the returns would be for much greater investment of resources. Indeed, the costs per positive result would increase exponentially (Evans, 1996, pp. 163–64). The prohibitive costs of such programmes have been a major deterrent to their introduction. However, nanotechnology might change this dramatically.

Currently, it has been estimated that it would cost 5 million GBP to map a complete genome (a lengthy process), and 250,000 GBP to map the 5 percent of the genome thought to be significant for medical purposes. Nanotechnology has been reported to promise vast reductions in both cost and time to the process, opening the door to rapid and affordable, large-scale genetic research. As Sheremeta (2004, pp. 50–51) has argued:

> Standard testing methods require large sample sizes and long reaction times to amplify the relevant genetic sequence using polymerase chain reaction (PCR). Microfluidic testing methods that are rapid and that can be performed on small biologic samples (for example, a single human cell) are currently being developed.

Sheremeta refers to published research papers that reveal that microfluidic chips can now be produced to perform PCR (Obeid et al., 2003, p. 288), which should make possible high through-put sequence analysis (Demello, 2003), and that it should be possible to read 2 million bases per second, thus enabling an entire genome to be sequenced in two hours.

All of this promise would make population screening of neonates, for example, affordable and achievable. Once the cost barriers are removed, it is probable that huge pressure will build up to introduce such programmes. Consequently, all the ethical and legal discussions which have proceeded at a relatively leisurely pace while specific genetic testing for very small minorities of patients have been provided will rapidly become intensified.

So what are these discussions? For the sake of convenience, we shall group the most notable areas of discussion under three headings. The first group of problems turns on the issue of consent, the second on the long-term benefits for neonates which, it has been claimed, would be afforded by such programmes, and the third on the nature of the knowledge gained by the screening programme and its consequences for the patients and their families.

1) Consent issues: If screening programmes were fully operationalized and of maximal usefulness, then all neonates should be screened. For example, the larger the number of neonates who are not included in the programme, the less likely it would be that researchers could draw generalized conclusions about relationships between genetic predispositions and social and demographic data. However, it is clear that there can be no ethical justification for obliging people to be involved either in research enterprises of this sort or in preventative or therapeutic interventions. This is clearly set out in Article 6 (a) and (b) of the *Universal Declaration* cited above, together with Article 6 (c):

> **Article 6** 3. In appropriate cases of research carried out on a group of persons or a community, additional agreement of the legal representatives of the group or community concerned may be sought. In no case should a collective community agreement or the consent of a community leader or other authority substitute for an individual's informed consent.

There are special problems relating to neonatal genetic screening which call for comment here, given the impossibility of obtaining consent from neonates. Parents are charged with giving or withholding consent for their children to take part in medical examinations and therapeutic interventions because they are thought, above all people, to have their children's interests at heart. It is not clear that they may consent to their children's participation in research, however, for such research cannot be said to be specifically in the children's interest. Aside from this difficulty, the creation of a genetic profile of a child at birth entails possibilities on which parents cannot possibly base their consent on behalf of the child, for how would the parents be able to determine what the child is likely to consider, in later life, what its best interest was at birth? Given the various burdens that knowledge of one's genetic make-up might impose, such as social stigma, restrictions of civil liberties in terms of obtaining insurance, employment and even education of certain sorts, many people choose not to be informed of genetic probabilities pertaining to them (Evans, 1999). As the neonate is in no position to resist such screening, the knowledge will be forced upon him or her, or those who care for him or her, and in later life since it will be part of his or her medical record.

Whenever genetic testing is offered to patients, genetic counselling is conducted in order to explain to them the consequences of being tested and the possibilities of positive results and their consequences. In light of such information, the decision is made on whether to be tested or not. Neonates cannot participate in the process at all.

While parental consent to prenatal genetic testing for conditions which could be developmentally important for the child is a proper responsibility of parents, screening for conditions which would, if at all, only develop during the neonate's adult life are another matter. They could be said to be options for the neonate on reaching maturity.

One of the protections provided to patients who enter clinical trials and who are involved in therapeutic interventions is the possibility of reviewing consent and withdrawing from the trial or refusing to continue treatment. While it is impossible to keep specific genetic knowledge from adults who have opted to receive it, it can at least be argued that the possession of such knowledge is voluntary. No such mitigation can be offered to a neonate who was genetically screened for a much wider range of possibilities.

Such negative considerations have to be weighed against the benefits promised for the individual neonate who is genetically screened. These might be taken into account when making a 'best interests' judgement on behalf of the incompetent child.

2) Promised benefits to the neonate: The UK Department of Health's White Paper (2003), to which the Human Genetics

Commission responded in their report on genetic profiling (HGC, 2005), referred to the analysis of a person's entire genome in order to reveal the majority of their genetic variations. The significance of only some of these variations is already known. As the neonate progresses into adulthood, no doubt more will become known about these matters. If those variations were noted in his or her medical records, the person concerned would stand to benefit from early warning of his or her propensity to develop certain conditions or possibly from learning of lifestyle changes which could help him or her to avoid triggering the development of the conditions to which he or she may be predisposed. This would appear to offer such individuals a considerable advantage. On the other hand, it is also likely that the evidence of causal links will not match the progress in identifying the predispositions to uncovered diseases. Indeed, this is currently true of marker genes which have already been identified. In such cases, the knowledge of one's predisposition to suffer the diseases might prove more of a burden than a blessing, except where it might extend one's freedoms. For example, while in some cases (notably regarding Huntingdon's Chorea) knowledge of the certainty that one will develop, or that one is a carrier of, the condition might inform reproductive decisions, in most cases the probabilistic knowledge provided by the possession of a marker gene will tend to limit one's civil liberties in ways canvassed earlier.

The development of pharmacogenomics also promises to promote the well-being of the neonates armed with genetic profiles. More medicines are being matched with groups of patients most likely to benefit from them, or least likely to suffer unacceptable adverse reactions to them, by means of genetic compatibilities. If their genetic profile has been determined by means of a neonatal screening programme, patients will later be spared the vagaries of trial-and-error prescriptions that would otherwise occur.

The usefulness of knowledge of a predisposition to a disease for which there is no known treatment has usually been referred to as disadvantageous, save in cases such as Huntingdon's. However, given the probable lifespan of the neonate, treatments which come on line will be more readily matched to them in the course of monitoring their health.

Some doubt remains about whether there is a positive effect on parents who possess knowledge of their child's genetic profile. While it would appear that such knowledge would alert the parents to dangers and thus enable them to provide better care to their children, there is evidence that knowledge of their child's predisposition to develop a disease tends to make them overprotective. This can have spin-off effects on the welfare of the child, who is cast in the patient role (Kerruish and Robinson, 2005).

The more general benefit thought to derive from neonatal genetic screening concerns the planning of future health provision. The more information planners can have about likely disease profiles in a society, the more intelligently they will be able to plan health service provision to match those developments. Since genetic predisposition is but one feature of disease development, it is not clear that this statistical data will be very useful in this regard. It might also militate against the interests of individual neonates who will be likely to develop rarer conditions than those for which explicit provision will be made, given the constant temptation of health service planners to maximize health gain from the investment of health funds. This raises serious questions about distributive justice in health care (Evans, 1993).

In addition, the possession of genetic profiles by patients might backfire on health planners and prove a hindrance to efficient provision of health services. The syndrome of the 'worried well' patient will likely occur. This can have a double effect. First, extra demands are placed upon the health service as a result of the anxieties that the genetic knowledge produces. A person who knows of a genetic susceptibility to a condition and develops an ache or pain that might conceivably be related to that condition will be likely to demand investigations in order to deal with the problem early. The number of false alarms will vastly outstrip the number of genuine symptoms and will place an enormous additional burdens on care-providers. Second, the anxiety produced by the knowledge might itself produce negative health consequences with which the health service will have to deal.

3) The nature of genetic data: There are ethical problems that arise from the very nature of genetic data. These are problems that have arisen in the use of genetic tests irrespective of the employment of nanotechnological aids. They also arise in neonatal screening programmes which in many countries are already mandatory. Nevertheless, the problems would be considerably intensified given the advent of the kind of neonatal screening outlined above.

There are two major areas of difficulty regarding the nature of genetic information. First, it is information that tells one a great deal not only about the presenting patient but about other people as well. This shared nature of genetic data raises the question of ownership. Is the neonate's genetic profile the neonate's property, and, if so, is it solely the neonate's property? If something is my property, I can determine whether and how others can make use of it. I have a right to use that property in any way I choose within the law and proscribe the use of it by others altogether, even if they have urgent need of it. If that property is shared, it will not be possible to control its use in this way, even if it used in ways that are against my interests.

When information contained in the profile of the neonate reveals something significant about the genetic makeup of a relative, does not that relative share the ownership of the information? And is not that relative entitled to know it? Consider the example of the child born with a severe genetic abnormality: the condition was the result of a carrier gene of the mother, whose sister was also likely to carry the gene. Should the sister be allowed access to this information in order to guide her reproductive choices? In the case in question, the mother refused permission for her sister to be told of the possibility because there were bad relations between the two and the mother hoped that her sister would suffer the same tragedy. What should the geneticist do in such situations? What about the child who has been discovered to be carrying the genetic mutation for Huntington's? Should the 25-year-old mother, whose father died of the disease at the age of 48, be told this even though she has made it clear that she does not want to know whether she has the mutation? And how does this affect the 45-year-old grandmother who has not expressed a wish not to know? Problems of this kind would become much more common were general screening programmes to provide the complete genome of neonates. Some guidance for practitioners or regulation of practice would be required upon the introduction of such a programme in order to protect both the neonate's and its relatives' interests, but they would not be easy to devise given the vast number of differences which will exist between cases (IBAC, 2002).

The second area of difficulty concerns another aspect of the aforementioned cases, namely privacy and confidentiality of this sensitive personal information. Where should we draw the line in defining the rights of other individuals and bodies to access the genetic profiles of neonates provided by a population screening programme? It is clear that open access would be detrimental to the child's interests, especially when he or she grows into adulthood and becomes subject to discrimination on the basis of the stigma attaching to the possibility of developing a variety of disease conditions or disabilities.

As we have seen, one of the motivations for the proposed introduction of such programmes is to enable governments to better engage in planning health services. Aside from the question of whether the probabilistic information contained in gene profiles could in fact provide bases for accurate prediction and planning, the further ethical question arises as to whether the information should be used for other purposes than those for which is ostensibly collected? The use of the Guthrie card offers us a salutary warning here. This card stores a blood sample from neonates which is obtained in order to test neonates for a series of serious but rare genetic disorders, including cystic fibrosis, congenital hypothyroidism and phenylketonuria. This test is provided because early intervention is in the interests of the children so afflicted. In many places, these tests are mandatory. The cards are kept as part of the children's medical record and so are stored in an identifiable form. It has been noted that on some occasions these cards have been used for purposes other than those for which the test was provided – for example, in police enquiries – and that in some places clinicians have expressed their willingness to release them to insurance companies, employment agencies and law enforcement agencies without subpoena (Elkin and Jones, 2000). The use of the data for health research also raises major privacy issues (Thomas, 2004).

The availability of complete genetic profiles would be an even more attractive source of information for both good and bad purposes for governments and other bodies of various kinds and would present the possibility of serious threats to civil liberties. For example, paternity testing for the enforcement of social service benefit provision and child allowance purposes would become a tempting possibility. It is not difficult to imagine much more serious threats, such social control measures in authoritarian settings. The information deriving from such screening programmes would therefore need to be very carefully controlled – assuming that such programmes should be allowed at all.

Nanotechnology and patient care

Many clinical interventions in the care of individual patients are being promised by nanotechnology. Most are no more than possibilities at present, but others are currently undergoing clinical trials. The pace at which such possibilities are being proposed is remarkable. For example, the following list of working possibilities were reported in the *New Scientist* between October 2004 and November 2005:

29 October 2004 Nanostructured contact lenses designed to release drugs into the eye, as needed, to treat conditions such as glaucoma.

27 November 2004 Use of gold nanoparticles to block angiogenesis, which is an important part of tumour development, and thus avoid having to destroy the results of such development in the treatment of cancer.

7 January 2005 Use of 'smart bombs' consisting of polymer capsules peppered with gold nanoparticles containing chemo drugs and attached to tumour seeking antibodies which will enter cancer cells and be exploded by low-energy laser pulses. This would kill the cells and leave surrounding cells untouched.

31 January 2005 Combinations of magnetic and gold nanoparticles with strands of DNA to enable early detection and possibly treatment of Alzheimer's disease.

2 April 2005 Use of nanosized laser beams able to detect one pre-cancerous cell amid other cells.

17 June 2005 Use of nanoparticles capable of delivering drugs selectively to cancer cells and thereby leaving surrounding cells unharmed.

11 September 2005 Creation of smart plastic films to coat implants, such as coronary stents and hip replacements, and slowly release drugs.

7 November 2005 The creation of new organs by using biodegradable nanotubes. They would contain either liver or kidney cells to create a vascular framework.

In addition, announcements have been made about the use of nanotechnology to treat spinal lesions in newly injured patients and the production of hearing implants which will make it possible to listen to music as well as speech. The list is growing by the day.

Clinical innovations

So what are we to say about all this? First, that it is extremely exciting for health care professionals and patients. The dream of operating at the cellular level seems within reach and will make current surgical techniques look crude. Choi (2004) has argued:

> Disease and ill health are caused largely by damage at the molecular and cellular level. Today's surgical tools are, at this scale, large and crude. It is said that from the viewpoint of a cell, even a fine scalpel is a blunt instrument more suited to tear and injure than heal and cure. Modern surgery works only because cells have a remarkable ability to regroup, bury their dead and heal over injury.

But is it safe to proceed with clinical trials at this point? This is the major question to be asked about these realistic applications, as well as about some of the more ambitious treatments with promise, such as the use of nanorobots called respirocytes, which will be 236 times more efficient at delivering oxygen to tissues than a natural red cell (Freitas, 1998).

Here we have to face the research finding that nanoparticles may cause blood clotting (ETC Group, 2002), as well as that they can accumulate in organisms (Brown, 2002). Thus far it is unclear how significant these findings are, even as we consider placing these particles into patients. The enthusiasm to help patients must not mask the need for adequate risk assessment regarding such innovative practices. Respect for Article 4 of the *Universal Declaration* is called for:

> **Article 4** In applying and advancing scientific knowledge, medical practice and associated technologies, direct and indirect benefits to patients, research participants and other affected individuals should be maximized and any possible harm to such individuals should be minimized.

We therefore have to take care that the introduction of nanotechnological techniques under the guise of innovative clinical treatment does not evade scrutiny by independent ethics committees. Some work has been done of late to provide guidelines for such review in the light of the conceptual difficulty of obtaining sound evidence for the effectiveness of such innovations by means of randomized controlled trials (Evans, 2002b; 2002a).

Assuming that the objective of each of these treatments is worthwhile, an ethics committee's first concern should be safety. It is unlikely that a committee would be satisfied with a theoretical account of the safety of procedures involving the introduction of nanoparticles into patients' bodies without a cohort of animal studies to examine the likely effect of such particles on blood flow, accumulation in organs, excretion of the particles in humans, etc. For example, the effectiveness of coatings mentioned earlier would need to be addressed. If satisfied that the treatments were acceptably safe as far as could be known, the committee would then have to consider how slowly the techniques would need to be introduced. Probably the most useful model would be that of drug development, where small numbers of patients would first be treated, subject to fully informed consent to test safety and then patients would be treated in the context of a randomized controlled trial. The technical difficulties related to blinding the studies can be ameliorated by securing independent assessment of patients and audit of results. Only when clinical equipoise disappears should such treatments be adopted as *bona fide* treatments, though even then they will need to be audited over the longer term, since exposure to nanoparticles might have long-term rather than short-term deleterious effects.

Cognitive and behavioural interventions

This particular category of potential clinical intervention calls for separate consideration because it raises important ethical issues about the identity of patients. There are two aspects to this identity problem: a) the question of the enhancement as opposed to the treatment of human beings and b) the question of authenticity of choices and behaviour.

What kinds of intervention give rise to these questions? We referred earlier to discussions of the implantation of nanocomputer-like particles in patients' brains to enhance memory and possibly to remedy deficiencies in a brain's functioning and in people's behaviour. These developments are probably more hypothetical than many of those touched on in the foregoing, but it is nevertheless worth exploring them briefly here, since some of the issues brought into bold relief by this kind of application are present in the relatively more mundane applications previously considered.

'Cyborg' (cybernetic organism) is a term coined to refer to beings that are part organism and part machine. It is especially applicable to the kind of interaction between machine and human organism which would be involved in the implantation of nanocomputers in the brain. Not only has it been suggested that vast extensions to memory might be achieved by these means but also behavioural modifications – for example, the control of criminal tendencies or, at a more detailed level, the control of the impulse to deceive.

Enhancement of human beings

If such a programme of interventions became possible, should it be subject to limits? Regarding genetic engineering, there has been much discussion of the difference between restorative therapeutic intervention and enhancement. The former would consist in restoring a person (or a foetus) to a normal state by rectifying abnormalities. The latter would consist in amplifying normal potential into extraordinary potential. These problems connected with such intervention and enhancement are intensified in the projected applications of nanotechnology.

Even therapeutic intervention has been the subject of discussion insofar as it assumes an understanding of what is normal which could undermine the dignity and interests of human beings. The identity of disability is a contestable issue. Some disability groups believe that it is society which is disabling rather than people who are disabled. In other words, disability is a social construct used to discriminate between and against people. Engagement in any kind of system, whether it be pre-implantation genetic screening or ante-natal genetic testing or interventions such as those currently under discussion, would be ethically suspect on this view (Wolbring, 2004). Thus, for example, the termination of 'damaged' and 'unrepairable' foetuses reflects on the value accorded to people in our society who live with the difficulties which cause these foetuses to be rejected.

Such difficulties aside, there is a separate concern about whether we should seek to engineer enlarged potential in human beings in excess of what they would have possessed, if healthy. Once again, it is necessary to clarify the ideas of norm and normality before we can determine what enhancement would be. We also have to accept that people are not born equal in terms of their potentialities, whether healthy or not. This latter point is not one of fairness. But when it comes to increasing the potential of prospective people, fairness does become an issue. It is a fact that we already engage in many kinds of enhancement post-birth, including education, which are not equally available to people. While some believe this to be unfair, others do not. However, we already have faced this issue in sport, where enhancement of performance by means of drugs, such as anabolic steroids, is ruled out on grounds of fairness rather than undue health risks to the athlete. Life is not a game, but people do have to compete to make their way in the world, and some are disadvantaged compared to others in their natural ability to do so. It is doubtful that the latter would be first in line as recipients of nano-enhancement technologies. Whatever other ethical objections one might raise, it is predictable that such technological enhancements would widen the gap between the privileged and underprivileged in our world.

Authentic choices and behaviours

The most serious group of ethical questions related to nano-interventions at this level have to do with the concept of the human being itself. At what point would we become worried that an intervention was in fact dehumanizing? Here, no doubt, there would be wide disagreement. How could such disagreement be handled?

Some have put the matter in terms of the question of whether cyborgs are persons. The question invites a 'yes or no' answer which, on reflection, is not easy to provide. We have some experience with approaches to this question from the wide-ranging discussions about the status of the early human embryo since the advent of IVF technologies. The most popular means employed philosophers to answer the question has been to propose criteria which have to be satisfied in order to identify an entity as a person. There is no unanimity on the issue. Consider, for example, some of the criteria which have been proposed. Persons must:

- have the concept of a continuous life (Tooley, 1972)
- be self-consciousness (Singer and Wells, 1984)
- be self-legislating (Engelhardt, 1996, p. 141)
- have intentional interests (Gillet, 1987)
- be a psychologically integrated unity capable of morally imputable actions (Bole, 1990)
- be capable of valuing one's own life (Harris, 1985, p. 7).

The problem with this representative selection of criteria is: 1) they are counterintuitive in that some of those we usually regard as persons do not satisfy them, and 2) according to some of the criteria, people who underwent nanotechnological improvement would be more morally considerable as persons after the intervention than they had been before it. Some have argued that our recognition of persons is not based on the application of such criteria at all (Evans, 2006).

This is not to say that any one of these criteria fails to mark out important issues that would inform our moral attitude towards nanotechnological alteration of cognitive and behavioural patterns. For example, if the interference with behaviour were of a certain sort – for instance, rendering one's behaviour involuntary, and possibly subject to control by others – we would consider the behaviour neither culpable nor admirable. However, we would deem such interference with free action reprehensible if not downright criminal.

Yet even here there will be differences between cases. The behaviours of patients are already subject to controls by technological means in ways which enhance their freedoms. For example, the use of pacemakers facilitate all the actions of their recipients, whose lives could not continue without them, and the use of shunts helps one avoid hydrocephalus and resulting brain damage. However, there is no interference with intention in such cases, cyborgs though the persons are insofar as they fit the definition of being a mix of organism and machine. Behaviour-modifying drugs, such as Ritalin in the treatment of Attention Deficit Hyperactivity Disorder, are presumed to enable – though they do have an effect upon intentional behaviour – recipients to behave authentically, intentionally and not impulsively, thus freeing them to be themselves. It is interesting that people can intend to take such medications in order to keep a job or complete a task and remain free to be their authentic selves without the drug when they so desire. Witty Ticky Ray (Sacks, 1985), the Tourette Syndrome patient, was such a case. It is debatable which of his selves was the authentic person. Was it the man who chose to take Haldol to enable him to lead a boring, flat life, working a weekday job? Or was it the creative, weekend drummer for whom the ticks did not matter?

How we react to someone who behaves towards us in an apparently generous or caring way, or in a hostile and aggressive manner, very much depends on what we know about the reasons for the behaviour. If we knew that their pacemaker facilitated the behaviour, this would make no difference at all to our moral reactions to them. If, however, we knew that the behaviour itself was engineered somehow and beyond the person's control or wishes, then our view would, of course, be very different. People can, no doubt, be turned into robots to varying degrees, and the more we would perceive this to be happening in nanotechnological interventions, the more morally repulsive those interventions would become. It is difficult to see, however, why we should have absolute objections to such technology, irrespective of these considerations, given the kinds of treatment already available and approved which have cognitive and behavioural consequences.

When the time comes to make policy decisions about such developments, care will have to be taken to respect Articles 3a and 5 of the *Universal Declaration on Bioethics and Human Rights*:

> **Article 3** 1. Human dignity, human rights and fundamental freedoms are to be fully respected.

> **Article 5** The autonomy of persons to make decisions, while taking responsibility for those decisions and respecting the autonomy of others, is to be respected. For persons who are not capable of exercising autonomy, special measures are to be taken to protect their rights and interests.

CONCLUSION

The development of nanotechnologies and their application to health provision give rise to remarkable, salutary possibilities. However, these outcomes cannot be pursued without due regard for the welfare and rights of all the stakeholders involved. The development of preceding technologies has shown that uncontrolled activity in pursuing worthy goals can have deleterious effects. For example, in the development of pharmaceuticals the well-being of research participants has at times been compromised in ways which have subsequently required proper regulation and ethical review. In such cases, including that of genetic engineering, development occurred prior to public ethical discussion of the possible benefits, harms and dangers. In the case of nanotechnologies, we have the opportunity to think ahead while the technological developments are at an embryonic stage. Ethical reflection by researchers, manufacturers, consumers and governments should help to ensure that unacceptable practices and outcomes are avoided and the most desirable outcomes are achieved.

BIBLIOGRAPHY

Bear, G. 2005. *Blood Music.* New York, Ibooks Science Fiction Classics.

Bole, T. J. 1990. Zygotes, souls, substances, and persons. *Journal of Medicine and Philosophy*, Vol. 15, No. 6, pp. 637–52.

Brown, D. 2002. Nano litterbugs? Experts see potential pollution problems. *Small Times*, 15 March. http://smalltimes.com/print_doc.cfm?doc_id=3266 or http://online.sfsu.edu/~rone/Nanotech/nano%20litterbugs.htm (Accessed 11 December 2006.)

Choi, K. 2004. Ethical issues of nanotechnology development in the Asia-Pacific region. P. Bergstrom (ed.), *Ethics in Asia-Pacific*. Bangkok, UNESCO Bangkok, pp. 327–67. http://unesdoc.unesco.org/images/0013/001359/135911e.pdf (Accessed 20 December 2006.)

Demello, A. J. 2003. DNA amplification moves on. *Nature*, Vol. 422, pp. 28–29.

Department of Health, UK. 2003. *Our Inheritance, Our Future: Realising the Potential of Genetics in the NHS*. Genetics White Paper, Cm 5791-II. London, Department of Health. http://www.dh.gov.uk/assetRoot/04/01/92/39/04019239.pdf (Accessed 20 December 2006.)

Drexler, E. 1986. *Engines of creation: The Coming Era of Nantechnology*. New York, Anchor Books/Doubleday.

Dunford, R., Salinaro, A., Cai, L., Serpone, N., Horikoshi, S., Hidaka, H. and Knowland, J. 1997. Chemical oxidation and DNA damage catalysed by inorganic sunscreen ingredients. *FEBS Letters*, Vol. 418, Nos. 1–2, pp. 87–90.

Elkin J. and Jones, D. G. 2000. Guthrie cards: Legal and ethical uses. *New Zealand Bioethics Journal*, Vol. 1, No. 2, pp. 22–26.

Engelhardt, H. 1996. *The Foundations of Bioethics*. 2d ed. Oxford, Oxford University Press.

ETC Group (Action Group on Erosion, Technology and Concentration). 2002. No small matter! Nanotech particles penetrate living cells and accumulate in animal organs. ETC Group, http://online.sfsu.edu%7Erone/Nanotech/nosmallmatter.html (Accessed 11 December 2006.)

Evans, D. 1993. Limits to care. D. Evans and Z. Szawarski (eds.), *Solidarity, Justice and Health Care Priorities*. Linköping, Sweden, University of Linköping Press, pp. 28–41.

———. 1996. The limits of health care. D. Greaves and H. Upton (eds.), *Philosophical Problems in Health Care*. Aldershot, England, Avebury, pp. 159–73.

———. 1999. Ethics and genetics. *New Zealand Medical Journal*, Vol. 112, pp. 109–12.

———. 2002a. Ethical issues in research, Part I: Research by stealth. *Science's Next Wave*. http://sciencecareers.sciencemag.org/career_development/previous_issues/articles/1750/ethical_issues_in_research_part_i_research_by_stealth/(parent) (Accessed 20 December 2006.)

———. 2002b. Ethical review of innovative treatment. *HealthCare Ethics Committee Forum*, Vol. 14, No. 1, pp. 53–63.

———. 2006. Qualifying as a person. S. Holm, M. Häyry and T. Takala (eds.), *Life of Value*. Amsterdam/New York, Rodopi (2007).

Foresight Nanotech Institute. 2006. *Supplying Clean Water Globally* (*Foresight Nanotech Update*, No. 56, Summer). http://foresight.org/publications/FNupdate56.pdf (Accessed 20 December 2006.)

Freitas, R. A. 1998. Nanomedicine FAQ. http://www.foresight.org/Nanomedicine/NanoMedFAQ.html (Accessed 12 October 2006.)

Gillet, G. R. 1987. Reply to J. M. Stanley: Fiddling and Clarity. *Journal of Medical Ethics*, Vol. 13, No. 1, pp. 23–25.

Harris, J. 1985. *The Value of Life*. London, Routledge.

Hoet, P. H. M., Nemmar, A. and Nemery, B. 2004. Health impact of nanomaterials? *Nature Biotechnology*, Vol. 22, No. 19, p. 19.

Human Genetics Commission (HGC). 2005. *Profiling the Newborn: A Prospective Gene Technology?* March. London, HGC. http://www.hgc.gov.uk/UploadDocs/Contents/Documents/Final%20Draft%20of%20Profiling%20Newborn%20Report%2003%2005.pdf (Accessed 20 December 2006.)

Independent Biotechnology Advisory Council (IBAC). 2002. *Genetic Testing: An Introduction to the Technology that is Changing Our Lives*. Wellington, New Zealand, IBAC.

Kerruish, N. J. and Robinson, S. P. 2005. Newborn screening: New developments, new dilemmas. *Journal of Medical Ethics*, Vol. 31, pp. 393–98.

Mertz, J. L. 2001. Technological and educational implication of nanotechnology infrastructure and educational needs. M. C. Roco and W. S. Bainbridge (eds.), *Societal Implications of Nanoscience and Nanotechnology*. NSET Workshop Report. Washington, DC, National Science Foundation, pp. 148–55.

Obeid, P. J., Christopoulos, T. K., Crabtree, H. J. and Backhouse, C. J. 2003. Microfabricated device for DNA and RNA amplification by continuous-flow polymerase chain reaction and reverse transcription-polymerase chain reaction with cycle number selection. *Analytical Chemistry*, Vol. 75, No. 2, pp. 288–95.

Oberdörster, E. 2004. Manufactured nanomaterials (Fullerenes, C60) induce oxidative stress in the brain of juvenile largemouth bass. *Environmental Health Perspectives*, Vol. 112, No. 10, pp. 1058–62.

Oberdörster, G., Sharp, Z., Atudorei, V., Elder, A., Gelein, R., Kreyling, W. and Cox, C. 2004. Translocation of inhaled ultrafine particles to the brain. *Inhalation Toxicology*, Vol. 16, Nos. 6–7, pp. 437–45.

Phoenix, C. and Drexler E. 2004. Safe exponential manufacturing. *Nanotechnology*, Vol. 15, pp. 869–72.

The Royal Society & The Royal Academy of Engineering. 2004. *Nanoscience and Nanotechnologies: Opportunities and Uncertainties*. 29 July. London, The Royal Society & The Royal Academy of Engineering. http://www.nanotec.org.uk/finalReport.htm (Accessed 22 December 2006.)

Sacks, O. 1985. *The Man Who Mistook His Wife for a Hat*. New York, Touchstone.

Scientific Committee on Cosmetic Products and Non-Food Products (SCCNFP). 2003. Opinion concerning zinc oxide (COLIPA No. S 76). SCCNFP/0649/03. Brussels, European Commission, Health & Consumer Protection DG. http://www.europa.eu.int/comm/health/ph_risk/committees/sccp/documents/out222_en.pdf (Accessed 22 December 2006.)

Sheremeta, L. 2004. Nanotechnology and the ethical conduct of research involving human subjects. *Health Law Review*, Vol. 12, No. 3, pp. 47–56.

Singer, P. A., Salamanca-Buentello, F. and Daar, A. 2005. Harnessing nanotechnology to improve global equity. *Issues in Science and Technology*, 9 July, pp. 57–64. http://www.issues.org/21.4/singer.html (Accessed 22 December 2006.)

Singer, P. and Wells, D. 1984. *Making Babies: The New Science and Ethics of Conception*. Oxford, Oxford University Press.

Thomas, C. 2004. Guthrie test samples: Is the problem solved? *New Zealand Bioethics Journal*, Vol. 5, No. 2, pp. 25–33.

Tooley, M. 1972. Abortion and infanticide. *Philosophy and Public Affairs*, Vol. 2, No. 1, pp. 37–65.

UNESCO. 2005. *Universal Declaration on Bioethics and Human Rights*. Paris. UNESCO.

Wolbring, G. 2004. Solutions follow perceptions: NBIC and the concept of health, medicine, disability and disease. *Health Law Review*, Vol. 12, No. 3, pp. 41–46.

Chapter 6
NANOTECHNOLOGY AND THE DEVELOPING WORLD

Erin B. Court, Fabio Salamanca-Buentello, Peter A. Singer and Abdallah S. Daar

The disparity between the standard of living in developed and developing nations and the socio-economic inequities within countries, constitute the most pressing ethical issues of our time (Benatar et al., 2005). People from developing countries have the same right to a fulfilled, healthy life as individuals from industrialized countries; however, in practice, this right has been eclipsed by hunger, poverty and disease. Global inequities are exacerbated by the fact that Science & Technology (S&T) and Research & Development (R&D) are disproportionately concentrated in industrialized nations and applied to their priorities, and the same holds true of nanotechnology. To address this disparity, it is necessary to engage policy-makers in developing nations to build knowledge societies to shape their own economic and social prospects (UNESCO, 2005). UNESCO can play a critical role in building such knowledge societies and in enabling nanotechnology to develop in such a way that it has a positive impact on less industrialized countries.

It has been widely recognized that S&T are critical components of development. The United Nations Development Program's *Human Development Report 2001* (UNDP, 2001) highlights the significant impact that S&T has had in reducing mortality rates and improving life expectancy between 1960 and 1990; and the 2005 report by the United Nations Millennium Project's Task Force on Science, Technology and Innovation (UNMP, 2005) addresses the essential role of technological innovation in achieving sustainable development.

WHICH APPLICATIONS OF NANOTECHNOLOGY ARE MOST RELEVANT TO DEVELOPING COUNTRIES?

At the McLaughlin-Rotman Centre for Global Health Program in the Life Sciences, Ethics and Policy, formerly knon as the Canadian Program for Genomics and Global Health (http://www.geneticsethics.net), we have researched in depth the ways in which emergent technologies such as biotechnology and nanotechnology can benefit the 5 billion people living in the developing world. As is the case with other technologies, advances in nanotechnology tend to be geared to the interests of industrialized countries. However, we have identified applications of nanotechnology that hold promise in responding to global challenges. Many developing countries are already engaged in nanotechnology research that is relevant to local needs.

In 2004 we conducted a study to identify and prioritize the applications of nanotechnology most likely to benefit developing countries in the 2003–2013 period (Salamanca-Buentello et al., 2005). We used a modified Delphi method, which is a structured and systematized process that employs a series of sequential questionnaires interspersed with controlled opinion feedback, to collect and distil knowledge from a panel of experts so as to build reliable group consensus on a specific judgement issue. The results of such a global foresight exercise can enable international S&T cooperation and can be used to address domestic needs of developing countries.

In our study, a panel of sixty-three experts was selected based on contacts identified in our previous survey of nanotechnology activity in developing countries (Court et al., 2004). A conscious effort was made to balance the panel with respect to gender specialty areas within nanotechnology and geographic distribution. Of the panellists, thirty-eight (60 percent) were from developing countries and twenty-five (40 percent) from developed countries; fifty-one panellists (81 percent) were male and twelve (19 percent) female.

For the purposes of this study, we derived our definition of nanotechnology from several sources. It was validated by experts in nanotechnology and in public policy and sustainable development:

> Nanotechnology is the study, design, creation, synthesis, manipulation, and application of functional materials, devices, and systems through control of matter at the nanometre scale (1–100 nanometres, one nanometre being equal to 1 × 10–9 of a meter), that is, at the atomic and molecular levels, and the exploitation of novel phenomena and properties of matter that usually appear at that scale.

Figure 6.1: Geographical distribution of the 63 panelists in 26 countries for the Delphi study on the top ten nanotechnologies that can potentially benefit developing countries

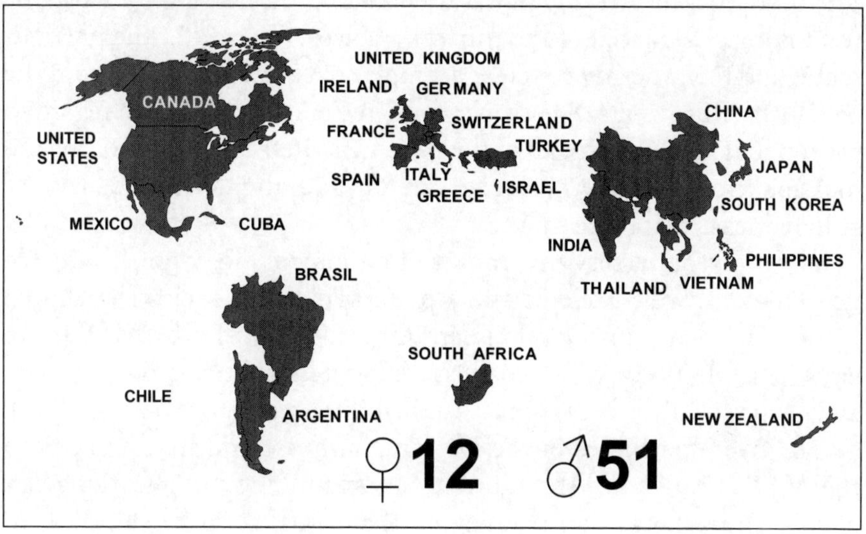

In the Delphi study we posed the following open-ended question: 'Which do you think are the nanotechnologies most likely to benefit developing countries in the areas of water, agriculture, nutrition, health, energy, and the environment in the next ten years?' These six areas had been identified in the 2002 World Summit on Sustainable Development in Johannesburg (UN, 2002). We asked the panellists to answer this question using the following criteria, which derived from a previous study in which we had identified the top ten biotechnologies for improving health in developing countries (Daar et al., 2002): impact, burden, appropriateness, feasibility and indirect benefits. Three Delphi rounds were conducted using e-mail messages, faxes and phone calls. In the first round, the panellists proposed examples of nanotechnologies in response to our study question. We analysed and organized their answers according to common themes and generated a list of twenty distinct nanotechnology applications. This list was reviewed for face and content validity by two nanotechnology experts who were not members of the Delphi panel. In the second round, the panellists ranked their top ten choices from the twenty applications provided and gave reasons for their choices. To analyse the data, we produced a summative point score for each application, ranked the list and summarized the panellists' reasons.

Then we redistributed the top thirteen applications, instead of the top ten, to generate a greater number of choices for increased accuracy in the last round. Thus the highest score possible for an application was 819 (63 × 13). The final round was devoted to consolidating consensus by re-ranking the top ten of the thirteen choices obtained in the previous round and to gathering concrete examples of each application from the panellists. Our results display a high degree of consensus with regard to the top four applications: all of the panellists cited at least one of the top four applications in their personal top four rankings, with the majority citing at least three (see table 6.1).

To further assess the impact of nanotechnology on sustainable development, we correlated the top ten applications with the United Nations Millennium Development Goals (MDGs). The MDGs are eight benchmarks by which to measure progress in human development and gauge social and economic sustainability (see Figure 6.2). In 2000, all 189 Member States of the United Nations committed to achieve the MDGs by 2015. The top ten nanotechnology applications can play an important role in achieving those goals (see Figure 6.3). In particular, examples of nanotechnologies identified by our panellists can be harnessed to address four key areas of sustainable development (Salamanca-Buentello et al., 2005; Court et al., 2005).

Developing renewable energy sources

One-third of the world's population relies primarily on traditional, non-renewable, contaminating fuels. Nanotechnology has the potential to provide developing countries with cleaner, more affordable and more reliable ways to harness renewable resources, averting recurrent energy crises, dependence on fossil fuels and environmental degradation brought about by the depletion of oil and coal. Solar cells, fuel cells and novel hydrogen storage systems based on nanostructured materials promise to deliver clean energy solutions. Quantum dots and ultrathin films of semiconducting polymers could significantly reduce the costs associated with conventional solar cells. Research on the photosensitization properties of nanoporous photovoltaic devices is currently being conducted in some developing nations. Electricity could also be produced cheaply by creating artificial systems that incorporate energy transduction proteins into an engineered matrix. Ideally, all of these applications would be robust and easily maintained and serviced.

Table 6.1: The top ten nanotechnologies with greatest potential to benefit the developing world

Ranking (score)	Applications of nanotechnology	Examples	Correlation with the MDGs
1	Energy storage, production and conversion	Novel hydrogen storage systems based on carbon nanotubes and other lightweight nanomaterials Photovoltaic cells and organic light-emitting devices based on quantum dots Carbon nanotubes in composite film coatings for solar cells Nanocatalysts for hydrogen generation Hybrid protein-polymer biomimetic membranes	VII
2	Agricultural productivity enhancement	Nanoporous zeolites for slow-release and efficient dosage of water and fertilizers for plants, and of nutrients and drugs for livestock Nanocapsules for herbicide delivery Nanosensors for soil quality and for plant health monitoring Nanomagnets for removal of soil contaminants	I, IV, V, VII
3	Water treatment and remediation	Nanomembranes for water purification, desalination and detoxification Nanosensors for the detection of contaminants and pathogens Nanoporous zeolites, nanoporous polymers and attapulgite clays for water purification Magnetic nanoparticles for water treatment and remediation TiO_2 nanoparticles for the catalytic degradation of water pollutants	I, IV, V, VII
4	Disease diagnosis and screening	Nanoliter systems (Lab-on-a-chip) Nanosensor arrays based on carbon nanotubes Quantum dots for disease diagnosis Magnetic nanoparticles as nanosensors Antibody-dendrimer conjugates for diagnosis of HIV-1 and cancer Nanowire and nanobelt nanosensors for disease diagnosis Nanoparticles as medical image enhancers	IV, V, VI
5	Drug delivery systems	Nanocapsules, liposomes, dendrimers, buckyballs, nanobiomagnets and attapulgite clays for slow and sustained drug release systems	IV, V, VI
6	Food processing and storage	Nanocomposites for plastic film coatings used in food packaging Antimicrobial nanoemulsions for applications in decontamination of food equipment, packaging or food Nanotechnology-based antigen detecting biosensors for identification of pathogen contamination	I, IV, V
7	Air pollution and remediation	TiO_2 nanoparticle-based photocatalytic degradation of air pollutants in self-cleaning systems Nanocatalysts for more efficient, cheaper, and better-controlled catalytic converters Nanosensors for detection of toxic materials and leaks Gas separation nanodevices	IV, V, VII
8	Construction	Nanomolecular structures to make asphalt and concrete more robust to water seepage Heat-resistant nanomaterials to block ultraviolet and infrared radiation Nanomaterials for cheaper and durable housing, surfaces, coatings, glues, concrete and heat and light exclusion Self-cleaning surfaces (e.g., windows, mirrors, toilets) with bioactive coatings	VII
9	Health monitoring	Nanotubes and nanoparticles for glucose, CO_2, and cholesterol sensors and for in-situ monitoring of homeostasis	IV, V, VI
10	Vector and pest detection and control	Nanosensors for pest detection Nanoparticles for new pesticides, insecticides and insect repellents	IV, V, VI

Source: Salamanca-Buentello et al., 2005.

FIGURE 6.2: The United Nations Millenium Development Goals

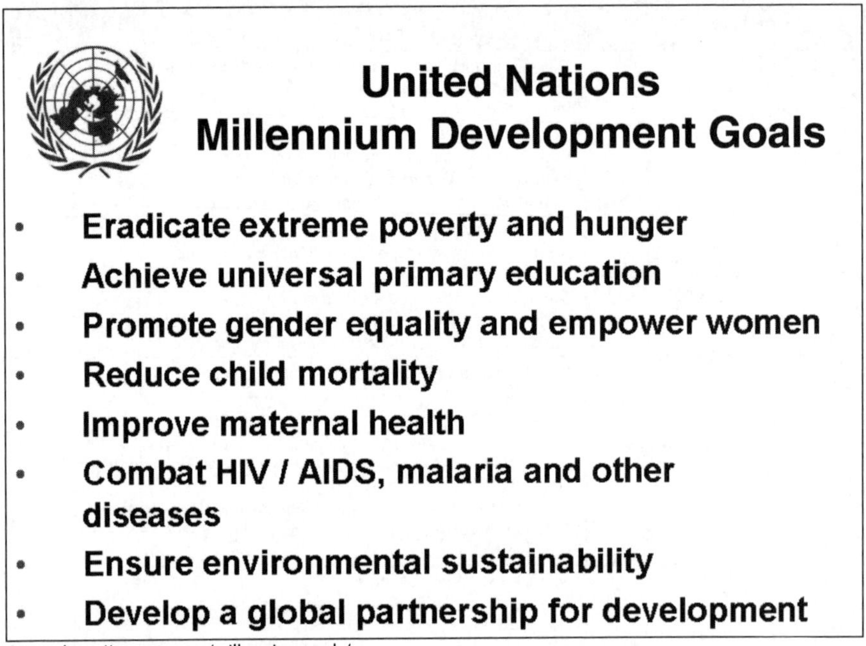

Source: http://www.un.org/millenniumgoals/

Figure 6.3: Correlation between the top ten nanotechnologies with the greatest potential to benefit developing countries and the MDGs

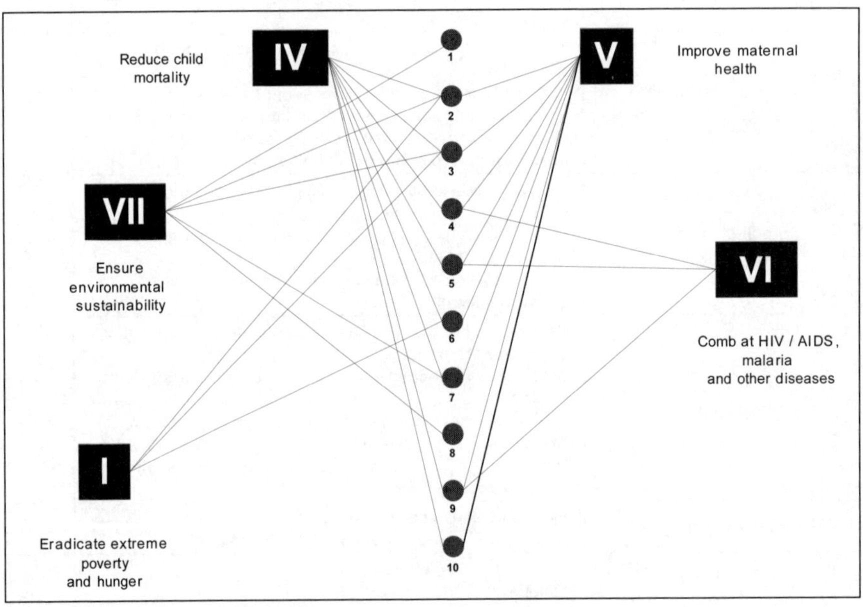

Source: Salamanca-Buentello et al., 2005

Promoting health

Nearly 3 million people in developing countries die of AIDS annually; tuberculosis (TB) accounts for more than a quarter of all preventable adult deaths; and in sub-Saharan Africa almost 1 million children die annually of malaria. Nanotechnology applications for diagnostic tools and drug and vaccine delivery are particularly promising for developing countries. Inexpensive, handheld, multiplex diagnostic kits could be used for wide-range screening in local clinics. Researchers are currently developing a prototype for a nanotechnology-based TB diagnostic kit. Microfluidic devices (lab-on-a-chip), biosensor arrays based on carbon nanotubes, magnetic nanoparticles and quantum dots offer significant advantages over conventional diagnostic methods. Nanotechnology could also be used in drug-delivery systems for the slow and targeted release of drugs, characteristics valuable for countries with no adequate drug storage capabilities or distribution networks. Nanotechnology could also potentially reduce transportation costs and even required dosages by improving the shelf-life, thermostability and resistance to changes in humidity of existing medications.

Reducing hunger

A substantial proportion of the population of developing countries lives in rural areas and does not have access to adequate nutritional sources. Malnutrition contributes to more than half of the deaths of children under five years of age in less industrialized nations. Several inexpensive agricultural applications of nanotechnology have the potential to decrease malnutrition, and thus childhood mortality, by increasing soil fertility and crop productivity. Nanoporous materials such as zeolites, which can form well-controlled stable suspensions with absorbed or adsorbed substances, could be employed for the slow release and efficient dosage of fertilizers for plants and of nutrients and drugs for livestock. Nanosensors could be used to monitor crop health and, when applied to the skin of livestock or sprayed on crops, could help detect the presence of pathogens. Moreover, nanotechnology-based methods of food packaging and storage may enable a wider and more efficient distribution of food products to remote areas in less industrialised countries.

Improving water and sanitation

One-sixth of the world's population lacks access to basic water supplies. More than one-third of the population of rural areas in Africa, Asia and Latin America has no clean water. More than 2 million children die each year from water-related diseases, such as diarrhoea, cholera, typhoid and schistosomiasis, which result from a lack of adequate water and sanitation. Arsenic, fluoride and nitrates threaten water supplies in many regions. Inexpensive, easily transportable and easily cleanable systems – such as nanomembranes and nanoclays – purify, detoxify and desalinate water more efficiently than conventional bacterial and viral filters. Nanoelectrocatalytic systems could be harnessed to decompose organic pollutants and remove salts and heavy metals from liquids, enabling the use of heavily contaminated and salinated water for drinking and irrigation. Nanomagnetic particles and nanoporous materials, such as zeolites and attapulgite, can also be used to absorb toxic heavy metals, organic pollutants and bacteria from water. Several of the contaminating substances retrieved could then easily be recycled.

WHAT ARE DEVELOPING COUNTRIES THEMSELVES DOING IN NANOTECHNOLOGY?

Developing countries should not be relegated to passivity, watching from the sidelines as industrialized countries profit from nanotechnology while negligible benefits trickle through to those in dire need. A focused policy on technology *innovation*, and not merely technology *transfer*, is requisite for addressing local needs. As *Forbes Magazine* publishes another year's 'Top ten nanotech products' featuring foot warmers, golf balls and personalized skin care (Wolfe, 2005), strategic consideration must be given to how less industrialized countries can develop their own nanotechnology innovation systems which are responsive to local priorities. A 2006 paper by Mohammed Hassan, President of the Academy of Sciences for the Developing World, shows the trend towards increasing investment in S&T in developing countries, and highlights those that are investing and making progress in nanotechnology (Hassan, 2005). Our previous survey of nanotechnology activity in developing countries (Court et al., 2004) revealed that many of these nations are engaged in a surprising amount of nanotechnology R&D activity (see Table 6.2). This activity takes the form of government-backed national nanotechnology initiatives, focused policies to encourage technological innovation, and local industries that have partnered with scientists and researchers to develop nanotechnology-related products to address local needs where they

Table 6.2: Different levels of nanotechnology activity in select developing countries

CATEGORY	COUNTRY	DEGREE OF NANOTECHNOLOGY (NT) ACTIVITY	EXAMPLES
Front runner	China South Korea India	• National NT government funding program • NT-related patents • Commercial products on the market or in development	**China:** • National Center for Nanoscience and Nanotechnology • Clinical trials of NT bone scaffold **South Korea:** • Nanotechnology Development Program • World's first carbon nanotube field emission display **India:** • Nanomaterials Science and Technology Initiative (NSTI) • Commercialization of nanoparticle drug deliveries
Middle ground	Thailand Philippines South Africa Brazil Chile	• Development of national NT government funding program • Some form of existing government support (e.g., research grants) • Limited industry involvement • Numerous research institutions	**Thailand:** • Center of Nanoscience and Nanotechnology, Mahidol University **Philippines:** • University of the Philippines/ Intel Technology Philippines optoelectronics project **South Africa:** • South African Nanotechnology Initiative (SANi) **Brazil:** • Institute of Nanoscience, Federal University of Minas Gerais **Chile:** • Nanotechnology Group, Pontificia Universidad Católica de Chile
Up and comer	Argentina Mexico	• Organized government funding not yet established • Industry not yet involved • Research groups funded through various S&T institutions	**Argentina:** • Nanoscience Research Group, Centro Atómico Bariloche and Instituto Balseiro **Mexico:** • Departamento de Materiales Avanzados, Instituto Potosino de Investigación Científica y Tecnológica

Source: Salamanca-Buentello et al., 2005

logically have a market. Below we show detailed country information acquired in our studies (Singer et al., 2006; Singer et al., 2005).

China

China has a very strong and solid Nanoscience and Nanotechnology National Plan, a National Steering Committee for Nanoscience and Nanotechnology, and a National Nanoscience Coordination Committee. Eleven institutes of the Chinese Academy of Sciences are involved in major nanotechnology research projects funded partly by the Knowledge Innovation Program. The Chinese Ministry of Science and Technology actively supports several nanoscience and nanotechnology initiatives. The Nanometre Technology Center in Beijing is part of China's plan to establish a national nanotechnology infrastructure and research centre; it involves recruiting scientists, protecting intellectual property rights, and building international cooperation in nanotechnology. At Tsinghua University, a group of scientists has developed an artificial bone called the 'nano bone', which has been used to treat eighteen patients with bone disease; according to scientists, the artificial implant is absorbed by the human body and replaced by human bone tissue. China's first nanometre technology industrial base is located in the Tianjin economic and development area. Haier, one of the country's largest home appliance producers, has incorporated a series of nanotechnology-derived materials and features in refrigerators, televisions and computers. Industry and academic researchers have worked together to produce nanocoatings for textiles that render silk, woollen and cotton clothing, water and oil proof; prevent clothing from shrinking; and protect silk from discolouration. Nanotech Port of Shenzhen is the largest manufacturer of single-walled and multi-walled carbon nanotubes in Asia. Shenzhen Chengying High-Tech produces nanostructured composite anti-ultraviolet powder, nanostructured composite photocatalyst powder and high-purity nanostructured titanium dioxide. The latter two nanomaterials are being used to catalyse the destruction of contaminants with the aid of sunlight.

India

Indian nanotechnology efforts cover a wide spectrum of areas, including microelectromechanical systems (MEMS), nanostructure synthesis and characterization, DNA chips, quantum computing electronics, carbon nanotubes, nanoparticles, nanocomposites and biomedical applications of nanotechnology. Through the Department of Science and Technology, the Indian Government catalysed the National Nanotechnology Program, which is funded with $10 million over three years. India has also created a Nanomaterials Science and Technology Initiative and a National Program on Smart Materials; the latter will receive $15 million over five years. This programme, which is focused on materials that respond quickly to environmental stimuli, is jointly sponsored by five government agencies and involves ten research centres. The Ministry of Defence is developing projects on nanostructured magnetic materials, thin films, magnetic sensors, nanomaterials and semiconductor materials. India has also formed a joint nanotechnology initiative with the European Union (EU). Several academic institutions are pursuing nanotechnology research and development, among them the Institute of Smart Materials Structures and Systems of the Indian Institute of Science; the Indian Institute of Technology; the Shanmugha Arts, Science, Technology, and Research Academy; the Saha Institute of Nuclear Physics; and the Universities of Delhi, Pune and Hyderabad. The Council for Scientific and Industrial Research, India's premier research and development body, holds numerous nanotechnology-related patents, including novel drug delivery systems, production of nanosized chemicals, and high temperature synthesis of nanosized titanium carbide. The University of Delhi obtained a US Patent in 2003 for nanoparticles for drug delivery. In the industrial sector, Nano Biotech Ltd. is doing research in nanotechnology for multiple diagnostic and therapeutic uses. This firm has developed immunodiagnostic tests for HIV, hepatitis B and syphilis; a nanodiagnostic kit for pregnancy has also been manufactured. Nano Biotech's targets for novel drug delivery systems include AIDS, infectious diseases and reproductive health. Dabur Research Foundation is involved in developing nanoparticle delivery systems for anticancer drugs. Similarly, Panacea Biotec has made advances in novel controlled-release systems, including nanoparticle drug delivery for eye diseases, mucoadhesive nanoparticles and transdermal drug delivery systems. CranesSci MEMS Lab, a privately funded research laboratory located at the Department of Mechanical Engineering of the Indian Institute of Science, is the first privately funded MEMS institution in India; it carries out product-driven research and creates intellectual property rights in MEMS and related fields with an emphasis on social obligations and education. Researchers at India's Banaras Hindu University, in collaboration with US researchers, have developed a method of large-scale production of carbon nanotube filters that could be used for water remediation. The India-based NGO Nimbkar Agricultural Research Institute (NARI) undertakes innovative S&T programmes to improve quality of life of the rural poor in India. It is currently investigating the use of nanoengines that operate on biomass-derived fuels to produce efficient and transportable lighting sources for rural communities. The group notes that once this technology is established, 'cost reduction processes and creative financing mechanisms for its availability to rural poor can be designed' (NARI, 2005).

Brazil

The government of Brazil considers nanotechnology a strategic area. The Brazilian National Nanotechnology Initiative started in 2001, combining several existing high-level nanotechnology research groups in several academic institutions and national research centres. Four research networks have been created with initial funds provided by the Ministry of Science and Technology through the National Council for Scientific and Technological Development. Two virtual institutes operating in the area of nanoscience and nanotechnology have also been created through the national programme. The total budget for nanoscience and nanotechnology for 2004 was about $7 million; the predicted budget for 2004–2007 is around $25 million. More than 400 scientists are working on nanotechnology in Brazil. There are research groups developing nanomagnets for cleaning water contaminated by large-scale oil spills; biodegradable micro- and nanoparticles that can potentially be used to deliver drugs selectively to cells and tissues; processes to generate diamond micro- and nanotubes of different porosities and with different kinds of tips for use in molecular 'cannons', implants, prostheses and luminescent nanodevices; nanotubes for biosensors and other electronic and mechanical nanocomponents of nanomachines; 3D meshworks for tissue bioengineering; computer remote-controlled robotic manipulators for minimally invasive surgeries and surgeries at a distance; myoelectric sensors (to determine the extent of muscular contractions), optic fibre sensors (to ascertain angles in movements), force sensors and temperature sensors; nanomagnets as drug transporters for medications against AIDS and cancer; and nanometric magnetic particles that are compatible with human blood and that possess physicochemical characteristics (physiological pH, salinity) which enable them to be injected without difficulty into the human bloodstream.

South Africa

South African research in nanotechnology currently focuses on applications for social development and industrial growth, including synthesis of nanoparticles, development of better and cheaper solar cells, highly active nanophase catalysts and electrocatalysts, nanomembrane technology for water purification and catalysis, fuel-cell development, synthesis of quantum dots, and nanocomposites development. The South African Nanotechnology Initiative (SANi), founded in 2003, aims to build a critical mass of universities, science councils and industrial companies that will focus on those areas of nanotechnology in which South Africa can have an advantage. To this end, SANi has an initial budget of about $1.3 million. The total spending in South Africa on nanotechnology is about $3 million. SANi is also interested in promoting public awareness of nanotechnology and assessing the impact of nanotechnology in the South African population. There are currently eleven universities, five research organisations (including the Water Research Commission) and ten private companies actively participating in this initiative. The areas of interest of the private sector in South Africa appear to be chemicals and fuels, energy and telecommunications, water, mining, paints and paper manufacturing.

Mexico

Mexico has thirteen centres and universities involved in nanotechnology research. In 2003, the National Council of Science and Technology spent $12.5 million on sixty-two projects in nineteen institutions. There is strong interest in nanoparticle research for optics, microelectronics, catalysis, hard coatings and medical electronics. Several groups have focused on fullerenes (in particular, carbon nanotubes), nanowires, molecular sieves for ultrahard coatings, catalysis, nanocomposites and nanoelectronics. Novel polymer nanocomposites are being developed for high-performance materials, controlled drug release, nanoscaffolds for regenerative medicine applications, and novel dental materials. In 2005, Mexican researchers, along with the Mexican federal government and private investors, unveiled a project for the creation of the $18 million National Laboratory of Nanotechnology, which will be under the aegis of the National Institute of Astrophysics, Optics, and Electronics. The initiative was funded by the National Council of Science and Technology, several state governments and Motorola.

WHAT ARE SOME OF THE CHALLENGES FACED BY DEVELOPING COUNTRIES IN NANOTECHNOLOGY?

Some have argued that nanotechnology will upset developing country export markets that rely heavily on raw materials, such as rubber, cotton, and agricultural products; the demand for these primary commodities will decrease as nanotechnology produces cheaper laboratory-created substitutes. It is contended that nanotechnology will displace poor mine, factory and agricultural labourers, among others (ETC Group, 2004). We disagree with this position. Advances in S&T have inevitably brought about the automation of manual tasks. Thus the decrease in manual labour is a natural consequence of any technology, not just nanotechnology. In both developed and developing countries, the workforce has necessarily adapted to the changes derived from successive waves of technology. In addition to providing the many benefits for developing countries detailed above, including the creation of other types of jobs, nanotechnology will create more opportunities for local scientists and researchers and will help to accelerate the transition of innovating developing countries towards robust knowledge-based economies. It is important to bear in mind that the economic impact of nanotechnology will vary depending on whether we are considering the more advanced developing countries, such as India and China, or the less developed.

Our previous case study of the health biotechnology innovation systems of seven developing countries (Thorsteinsdottir et al., 2004) identified private-sector involvement as a key factor in moving from knowledge to product. Nanotechnology will have an impact on all sectors of the economy. Those individuals, industries or nations that adapt to the changes brought about by nanotechnology and that engage in R&D to innovate in this field will enjoy the benefits of this new technology; those individuals, industries and nations that do not keep up with the changes will find it difficult to survive in a rapidly changing environment, one that is moving more and more towards economies based on knowledge (see Enriquez, 2001).

WHICH SAFETY, REGULATORY AND PUBLIC ENGAGEMENT ISSUES NEED TO BE ADDRESSED?

Effects on environment and human health

While there is clear evidence that nanotechnology can contribute to human welfare, particularly in developing countries, not all of the effects of nanomaterials in the environment are currently known. It is possible that nanoparticles could accumulate in different organisms, become incorporated in living tissue, and move up the food chain in a process known as bioaccumulation. Animal studies have found that inhaled nanoparticles can be selectively transported into the brains of mammals via the olfactory nerves. Additionally, inflammatory lesions can be induced by nanotubes in the lungs of mice. A study on large-mouth bass found elevated levels of oxidative deterioration of lipids in the brains of fish that were exposed to water-soluble fullerenes. While these studies suggest that nanomaterials may be toxic, animal models alone are not sufficient to predict their effects in humans. The details of the risks are beyond the scope of this chapter. (A useful review can be found in Theodore and Kunz, 2005.) It should be noted, however, that even as studies of the risks are being carried out, an increasing number of consumer and industrial products are already entering the market (see, e.g., the list in Wolfe, 2005, and announcements of even more products at Project on Emerging Nanotechnologies, 2006).

Regulatory issues

It is critical to develop regulatory frameworks, either by building upon existing regimes or by creating entirely new ones, that take into account the novel properties of matter at the nanoscale (see Davies, 2006). Given that regulatory reform is necessary, governments in both developed and developing countries ought to ensure that new regulatory frameworks are logical, efficient, transparent and readily adaptable to fast-paced technological change. At the international level, clear standards for workplace exposure to nanomaterials must be set. Additionally, international definitions and standards for measurement of the concentration and toxicity of nanomaterials in the environment must be developed. This should be carried out by independent bodies that remain under public scrutiny. Specific attention also needs to be directed towards recognizing individual responsibility of scientists and engineers who are responsible for generating advances in nanotechnology. The best way to prevent the misuse or abuse of nanotechnology may involve bringing together large groups of diverse stakeholders so that the areas where controversial issues may arise can be identified at the earliest possible stage. A role for benefit-sharing between developed and developing countries in the commercialization of nanotechnology products should be established. Consideration must be given to the best ways to harness intellectual property (IP) for the development of nanotechnology, bearing in mind that the inappropriate use of patents from universities, companies and the military might inhibit R&D efforts in less industrialized countries. There is an inherent tension with IP rights between providing incentives and rewards for innovation on the one hand, and promoting wide and easy access to the resultant technologies on the other hand, especially if the technologies are important to save lives or are needed by the majority. While such issues are complex, it seems prudent to avoid undue concentration of resources in a few hands, for this leads to 'patent thickets' which may actually impede R&D (Sabety, 2004). Alternatives are being studied to avoid these problems, including patent pools, patent clearinghouses and open source approaches, which have been successful in the IT sector. In the area of technology transfer, the African Agricultural Technology Foundation (AATF) has negotiated royalty-free licensing and sub-licensing with technology patent holders from private companies, public private partnerships and NGOs to help developing country research institutes to access patented agricultural technologies (see AATF, 2002).

Public engagement

Public involvement in crucial decisions about the use of nanotechnology is not optional but essential. However, the public cannot make informed decisions if it knows very little about nanotechnology. Thus public education about nanotechnology is absolutely necessary. We encourage candid discussion of the risks and benefits which is receptive to public concerns. Public debate should include discussions on control of nanotechnology, access to trustworthy information, the terms according to which nanotechnology is introduced into society, the benefits of nanotechnology, the risks of nanotechnology, the groups that will either benefit or be at risk, and the definition of the individual(s) who will take responsibility for problems that arise through research, development, commercialization and use of nanotechnology applications. The public backlash against genetically modified organisms reveals that public perception of a technology, no matter how detached from scientific evidence, will determine the public's approval or rejection of this technology. Thus, although it is very difficult to do well, it is in the best interest of all stakeholders to promote a well-informed public that can make objective and rational decisions. Public opinion, once entrenched, is very difficult to change. The more understanding the public has about nanotechnology, the better prepared it will be to make rational assessments instead of relying on knee-jerk reactions and gut feelings. Addressing the legitimate concerns and permitting developing countries to weigh the risks and benefits for themselves will allow for the responsible development of nanotechnology and will pave the way towards public acceptance. Funding of social science research in nanotechnology can be a good way to lend transparency to the debates on nanotechnology, since social scientists can add richness to the discussion on this topic. They are positioned to extend the discussions to the public and policy-makers while respecting the professional integrity of scientists working in the field. In fact, the input of social scientists can maximize the societal benefits of the technology while reducing the possibility of debilitating public controversies.

Engaging youth in discussions of emerging technologies is vital to developing an informed future voting population. In 2002, we at the then-called Canadian Program on Genomics and Global Health developed 'Engage: Stem Cells', a curricular unit for Canadian high school students to debate the ethical and governance issues of stem cell technology (http://www.stemcellnetwork.ca/engage). Based on this highly successful strategy, we are currently developing 'WaterEngage', a public outreach tool to foster greater social awareness about the potential of nanotechnology and biotechnology to improve lives in developing countries through the application of these technologies to water treatment and remediation.

The former Secretary-General of the United Nations, Kofi Annan, noted in 2003 that 'if every nation gains full access to [the] broader world community of science and has the opportunity to develop an independent science capability, its public can engage in a candid dialogue about the benefits and risks of new technologies, such as genetically engineered organisms or nanotechnology, so that informed decisions can be made about their introduction into our lives' (Annan, 2003).

Recent studies on the public perception of emerging nanotechnologies (Macoubrie, 2005; Scheufele and Lewenstein, 2005) suggest that the public desires more information on nanotechnology's promises and risks, and wants increased testing on environmental and health effects. The studies encourage industry to heed the public's message and recommend mechanisms to integrate the public in national policy decision-making. It is necessary to avoid 'marketing' campaigns in which the benefits of nanotechnology products are exaggerated while the risks are swept underneath a carpet of silence. One way to increase public participation in S&T is to convene citizens' juries, representative of a diverse subset of the population, which then provide recommendations for the future direction and regulation of nanotechnology. An example of this strategy is Nano Jury UK (http://www.nanojury.org), in which twenty randomly chosen people were brought together to take part in the discussion of the social impacts of nanotechnology and to provide a potential vehicle for the public's views to influence policy decisions.

WHAT CAN INDUSTRIALIZED COUNTRIES DO TO ASSIST DEVELOPING COUNTRIES IN HARNESSING NANOTECHNOLOGY?

Global challenges

One way to accelerate the application of nanotechnology to meet the needs of developing countries consists in summoning the world's best scientific minds to address grand challenges, using the model employed by the Foundation for the NIH / Bill and Melinda Gates Foundation's Grand Challenges in Global Health (Varmus et al., 2003). A Grand Challenge is a call to arms for investigators to channel their efforts towards a specific scientific or technological breakthrough that will overcome one or more significant development challenges. Grand Challenges identify obstacles between where we are and where we want to be. They permit multiple actors to focus not only on one particular problem but on the hurdles that need to be overcome. An initiative entitled 'Addressing Global Challenges Using Nanotechnology' (Salamanca-Buentello et al., 2005; Singer et al., 2005) could draw from the top ten nanotechnology applications we have already identified for developing countries to construct a plan of action for mobilizing the international community.

Table 6.3: Grand Challenges in Global Health projects that include a nanotechnology component

Grand challenge #2: prepare vaccines that do not require refrigeration
Bacterial spores as vaccine delivery systems
Thermostable vaccines with improved stability at non-refrigerated temperatures
Grand Challenge #3: Develop needle-free delivery systems for vaccines
Development of a targeted mucosal vaccine delivery technology
Nanoemulsions as adjuvants for nasal-spray vaccines
Needle-free delivery of stable, respirable powder vaccine
Needle-free vaccination via nanoparticle aerosols
Surface modified nanostructures as delivery vehicles for transmucosal vaccination
Grand Challenge #14: Develop diagnostic tools for multiple conditions at point of care
A point-of-care diagnostic system for the developing world

Source: http://www.grandchallengesgh.org/

Helping to secure funding

The funding to address global challenges using nanotechnology could come from various sources, including national and international foundations, from collaboration among nanotechnology initiatives in industrialized and developing countries, and from public and private sources. In February 2004, former Canadian Prime Minister Paul Martin proposed that 5 percent of Canada's R&D investment be used to address developing world challenges (Government of Canada, Office of the Prime Minister, 2004). If all industrialized nations adopted this target, part of these funds could be directed towards addressing global challenges using nanotechnology. In addition, governments of developed countries could provide incentives for their companies to direct a portion of their R&D towards the development of nanotechnology in less industrialized nations. In parallel to the allocation of public funds, policies should provide incentives to the private sector to direct a portion of their research and development towards funding the Grand Challenges initiative.

Forming effective North-South collaborations

There are already promising examples of North-South partnerships. For instance, in 2005 the EU allocated 285 million Euros through its 6th Framework Programme (FP6) for scientific and technological cooperation with third-partner countries, including Argentina, Chile, China, India and South Africa. A priority research area under FP6 is nanotechnology and nanoscience, to which 1,429 million Euros have been allocated (Cordis, 2002). Another example is the US funding of nanotechnology research in Vietnam, as well as the US-Vietnam Joint Committee for Science & Technology Cooperation. IndiaNano, a platform created jointly by the Indian-American community in the Silicon Valley and Indian experts involved in nanotechnology R&D, aims to establish partnerships between Indian academic, corporate, government and private institutions to support nanotechnology R&D in India and to coordinate the academic, government and corporate sectors with entrepreneurs, early-stage companies, investors, joint ventures, service providers, start-up ventures and strategic alliances (Singer et al., 2006; Singer et al., 2005).

Forming effective South-South collaborations

The successful experiences of several developing countries, such as China, India and Brazil, in harnessing nanotechnology to address some of their most pressing needs should be taken into account by other less industrialized nations. Some of the possible factors leading to this success include: focus on the use of nanotechnology to deal with local health needs; political will; long-term government support (in particular, development of specific polices, provision of funding, recognition of the importance of nanotechnology R&D, focus on the challenges of brain drain, and incentives for local nanotechnology enterprises to help them overcome problematic economic conditions); strong collaboration and linkages between academia, government and industry; definition of niche areas; use of competitive advantages; and private-sector development (Thorsteinsdottir et al., 2004). The UN Commission on Private Sector and Development report, *Unleashing Entrepreneurship: Making Business Work for the Poor* (CSPD, 2004), underscores the importance of partnerships with the private sector, especially the domestic private sectors in developing countries, in working to achieve the MDGs.

Facilitating knowledge repatriation by diasporas

We have recently begun an in-depth empirical study of how the scientific diaspora working in developed countries can more systematically contribute to innovation and development in their countries of origin (Séguin et al., 2003). A diaspora in this context is a community of individuals from a specific developing country who left home to attend school or find a better job and who now work in industrialized nations in academia, research or industry. This movement of highly educated men and women is often described as a 'brain drain' and is usually seen as having devastating effects in the developing world. Rather than deem this migration (which is extremely difficult to reverse) an unmitigated disaster, some developing countries have sought ways to tap these emigrants' scientific, technological, networking, management and investment capabilities. India actively encourages its 'non-resident Indian' diaspora to make such contributions to development back home, and these people have made a valuable contribution to the Indian ITC sector. We foresee a significant role for diasporas in the development of nanotechnology in less industrialized nations.

Global governance

A global strategy is needed to assess and promote nanotechnology for development. While traditional governance models tend to focus on risks and restrictions, a more holistic model should include a focus on innovation and the enormous potential of nanotechnology to address global challenges. Among the issues to be addressed are: how can nanotechnology be used responsibly? Which applications ought to be given priority? What policies are needed to develop capacity and ensure that the benefits of nanotechnology reach those in greatest need? How ought regulatory regimes to be developed and implemented? Recently several organisations have shown an increased interest in answering these questions (Dowdeswell et al., 2006). The most salient initiatives include those spearheaded by the Academy of Sciences for the Developing World (formerly the Third World Academy of Sciences; see http://www.twas.org) and its president, Mohamed Hassan; the Meridian Institute's Global Dialogue on Nanotechnology and the Poor (GNDP; http://www.meridian-nano.org/gdnp.php); and the Foresight Institute's 'Foresight Nanotechnology Challenges' (1986). Thought should be given to what governance model is most appropriate for mobilizing international action and institutionalizing cooperation in nanotechnology. International treaties and conventions may not be well suited to promoting the global benefits of a rapidly advancing field of S&T. They are often slow to be negotiated and ratified, and may still suffer from poor adherence once they come into force. A flexible global network of representatives from government, academia, industry and civil society could serve as a forum for international dialogue between the public and policy-makers to shape institutional decisions and discuss policy options. Equality of participation through engagement of developing country voices and the inclusion of diverse sectors like civil society and industry would facilitate collaborative strategies for managing risks in order to promote the global benefits of nanotechnology. Existing networks, public-private partnerships and coalitions working to establish worldwide capacity in S&T, including the InterAcademy Council (www.interacademycouncil.net) and The New Partnership for Africa's Development (NEPAD; www.nepad.org), could play a significant role in this global network.

CONCLUSION

In summary, both developed and developing countries must do their part to ensure that advances in nanotechnology lead to solutions for humanity's most significant hurdles. The key feature consists in addressing sustainable development issues by focusing on S&T innovation. In a 2002 speech to the National Academy of Sciences, former Secretary of State Colin Powell noted: 'You don't have to . . . be Secretary of State to survey the 21st-century terrain and see that science and technology must inform and support our foreign policy-making in this challenging world that we live in. Whether the mission is . . . creating conditions for sustainable development, or stemming the global HIV/AIDS pandemic, the formulation of our foreign policy must proceed from a solid scientific foundation' (Powell, 2002).

We have previously argued that industrialized countries should undertake the responsibility of 'genome diplomacy' to ensure that the benefits of the genomics revolution are extended to populations in developing countries (Daar et al., 2003). By analogy, we have put forth the concept of 'nanodiplomacy' (Salamanca-Buentello et al., 2005). Both developed and developing nations must leverage their national nanotechnology assets into foreign policy and, simultaneously, marshal both public and private funds to assist the developing world to address sustainable development challenges.

BIBLIOGRAPHY

African Agricultural Technology Foundation (AATF). 2002. Rationale and design of the AATF. Nairobi, Kenya, AATF. http://www.aatf-africa.org/rationale.php (Accessed 20 December 2006).

Annan, K. 2003. A challenge to the world's scientists. *Science*, Vol. 299, p. 1485.

Benatar, S. R., Daar, A. S. and Singer, P. A. 2005. Global health challenges: The need for an expanded discourse on bioethics. *PLoS Medicine*, Vol. 2, No. 7, e143, pp. 0587–89.

Community Research & Development Information Service (Cordis). 2002. *Integrating and Strengthening the European Research Area (2002–2006).* Sixth Framework Programme. Brussels, Cordis. http://cordis.europa.eu/nmp/home.html and ftp://ftp.cordis.europa.eu/pub/documents_r5/natdir0000066/s_1883005_20040723_094518_1883en.pdf (Accessed 20 December 2006.)

Court, E. B., Daar, A. S., Martin, E., Acharya, T. and Singer, P. A. 2004. Will Prince Charles et al diminish the opportunities of developing countries in nanotechnology? *Nanotechnology and Society*, Vol. 4, No. 3. http://www.nanotechweb.org/articles/society/3/1/1/1 (Accessed 20 December 2006.)

Court, E. B., Daar, A. S., Persad, D. L., Salamanca-Buentello, F. and Singer, P. A. 2005. Tiny technologies for the global good. *Nano Today*, May, pp. 14–15.

Daar, A. S., Dowdeswell, E. and Singer, P. A. 2003. Genome diplomacy: Canada's crucial role. *Policy Options*, Vol. 24, pp. 56–61.

Daar, A. S., Thorsteinsdóttir, H., Martin, D., Smith, A. C., Nast, S. and Singer, P. 2002. Top ten biotechnologies for improving health in developing countries. *Nature Genetics*, Vol. 23, pp. 229–32.

Davies, J. C. 2006. *Managing the Effects of Nanotechnology.* Washington, DC, Project on Emerging Nanotechnologies. http://www.wilsoncenter.org/index.cfm?topic_id=166192&fuseaction=topics.item&news_id=165552 (Accessed 20 December 2006.)

Dowdeswell, E., Daar, A. S., Acharya, T. and Singer, P. A. 2006. Realizing the promise of genomics: Exploring governance. *International Journal of Technology and Globalization* (in press).

Enriquez, J. 2001. *As the Future Catches You*. New York, Three Rivers Press.

ETC Group. 2004. *Down on the Farm: The Impact of Nano-scale Technologies on Food and Agriculture*. Winnipeg, ETC Group.

Foresight Nanotech Institute. 1986. Foresight nanotechnology challenges. Palo Alto, CA, Foresight Institute. http://www.foresight.org/challenges (Accessed 20 December 2006.)

Government of Canada, Office of the Prime Minister (2004) Complete text and videos of the Prime Minister's reply to the speech from the throne. http://www.pm.gc.ca/eng/news.asp?id=277 (Accessed 27 January 2005.)

Hassan, M. H. A. 2005. Nanotechnology: Small things and big changes in the developing world. *Science*, Vol. 309, pp. 65–66.

Macoubrie, J. 2005. *Informed Public Perceptions of Nanotechnology and Trust in Government*. Washington, DC, Project on Emerging Nanotechnologies. http://www.wilsoncenter.org/events/docs/macoubriereport.pdf (Accessed 20 December 2006.)

Nimbkar Agricultural Research Institute (NARI). 2005. Rocket science for rural development. http://nariphaltan.virtualave.net/rocketscience.pdf (Accessed 20 December 2006.)

Powell, C. 2002. Remarks to the 139th Annual Meeting of the National Academy of Sciences. 30 April. Washington, DC, National Academy of Sciences. http://www.nasonline.org/site/PageServer?pagename=NEWS_address_04302002_Powell_2002am (Accessed 20 December 2006.)

Project on Emerging Nanotechnologies. 2006. A nanotechnology consumer products inventory. Washington, DC, Project on Emerging Nanotechnologies. http://www.nanotechproject.org/index.php?id=44 (Accessed 20 December 2006.)

Sabety, T. Nanotechnology innovation and the patent thicket: Which IP policies promote growth? Abstract of paper presented at the First Conference on Advanced Nanotechnology: Research, Applications, and Policy, 21–24 October 2004. Palo Alto, CA, Foresight Nanotech Institute. http://www.foresight.org/Conferences/AdvNano2004/Abstracts/Sabety (Accessed 20 December 2006.)

Salamanca-Buentello, F., Persad, D. L., Court, E. B., Martin, D. K., Daar, A. S. and Singer, P. A. 2005. Nanotechnology and the developing world. *PLoS Medicine*, Vol. 2, No. 4, e97, pp. 0300–3. http://www.utoronto.ca/jcb/home/documents/PLoS_nanotech.pdf (Accessed 20 December 2006.)

Scheufele, D. A. and Lewenstein, B. V. 2005. The public and nanotechnology: How citizens make sense of emerging technologies. *Journal of Nanoparticle Research*, Vol. 7, No. 6, pp. 659–67.

Séguin, B., Singer, P. A. and Daar, A. S. 2003. Scientific diasporas. *Science*, Vol. 312, pp. 1602–3.

Singer, P. A., Daar, A. S., Salamanca-Buentello, F. and Court, E. B. 2006. Nano-diplomacy. *Georgetown Journal of International Affairs*, Winter, pp. 129–37. http://www.utoronto.ca/jcb/about/publications/nano-diplomacy.pdf (Accessed 20 December 2006.)

Singer, P. A., Salamanca-Buentello, F. and Daar, A. S. 2005. Harnessing nanotechnology to improve global equity. *Issues in Science and Technology*, 9 July, pp. 57–64.

Theodore, L. and Kunz, R. G. 2005. *Nanotechnology: Environmental Implications and Solutions*. Hoboken, NJ, Wiley.

Thorsteinsdottir, H., Quach, U., Daar, A. S. and Singer, P. A. 2004. Conclusions: Promoting biotechnology innovation in developing countries. *Nature Biotechnology*, Vol. 22 Suppl, pp. DC48–52.

UN Millennium Project, Task Force on Science, Technology and Innovation. 2005. *Innovation: Applying Knowledge in Development*. New York, UN Millennium Project. http://www.unmillenniumproject.org/reports/tf_science.htm (Accessed 20 December 2006.)

United Nations. 2002. World Summit on Sustainable Development, 26 August–4 September 2000 in Johannesburg, South Africa. http://www.un.org/jsummit/html/documents/summit_docs.html (Accessed 20 December 2006.)

UN Commission on Private Sector and Development (CPSD). 2004. *Unleashing Entrepreneurship: Making Business Work for the Poor*. New York, UN Commission on Private Sector and Development. http://www.undp.org/cpsd/documents/report/english/fullreport.pdf (Accessed 20 December 2006.)

United Nations Development Program (UNDP). 2001. *Human Development Report 2001: Making New Technologies Work for*

Human Development. New York, UNDP. http://hdr.undp.org/reports/global/2001/en (Accessed 20 December 2006.)

UNESCO. 2005. *UNESCO World Report: Towards Knowledge Societies*. Paris, UNESCO. http://unesdoc.unesco.org/images/0014/001418/141843e.pdf (Accessed 20 December 2006.)

Varmus, H., Klausner, R., Zerhouni, E., Acharya, T., Daar, A. S. and Singer, P. A. 2003. Grand challenges in global health. *Science*, Vol. 302, pp. 398–99. http://www.grandchallengesgh.org (Accessed 20 December 2006.)

Wolfe, J. 2005. Top ten nanotech products. *Forbes/Wolfe Nanotech Report*, 12 January. http://www.forbes.com/investmentnewsletters/2005/01/12/cz_jw_0112soapbox.html (Accessed 20 December 2006.)

Chapter 7

PUBLIC ENGAGEMENT AND EDUCATION FOR ETHICS IN NANOTECHNOLOGY

Kyunghee Choi

Major industrialized countries are incorporating nanotechnology in their innovation systems as a means to generate wealth in the near future. Over thirty national governments have launched nanoscience initiatives, and begun to invest heavily in R&D (Roco, 2003). Expenditures have approximately tripled between 1997 and 2002 in Western Europe and many countries in Asia and the Pacific, such as China, South Korea and Australia, have begun investing significant sums in nanotechnology.

Needless to say, nanotechnology will enrich many facets of human life. The move into the nanoworld implies virtually absolute control over matter at the level of the single atom or molecule. Advanced nanotechnology can build machines thousands of times more powerful, and hundreds of times cheaper, than today's devices.

Reflection on new technologies is indispensable for their use since it helps one to avoid unexpected or unintended negative consequences. In terms of controlling the malicious uses of technology, it was only recently that educators and scientists put the related ethical issues on the agenda. In relation to nanotechnology, these issues are perceived with different weight based on the nation's cultural background, as well as their people's belief systems, tradition and developmental growth.

In seeking to identify the ethical issues pertaining to nanotechnology, it becomes evident that they overlap with the ethical issues involved in other technologies, since nanotechnology – including nanobiotechnology, nano-information technology, and nano-environmental technology – is an applied technology. This makes its ethical issues even more complex and comprehensive. In addition to all the potential benefits of technology, there is a flip side, which concerns how the interactions of these technologies affect everyday life in the future. Future initiatives will continue to be explored in order to gain a better understanding of the ethical implications of nanotechnology (Roco and Tomellini, 2002).

Many potential negative effects of nanotechnology have been identified and become the subject of debate. For example, the National Science Foundation (NSF) in the US has issued proceedings and reports on the social implications of nanoscience and nanotechnology. In the European Union, discussions of the ethical implications of nanotechnology have also started, for example in Germany and Denmark. Having recognized the complexity of the ethical agenda regarding nanotechnology, some countries have started to educate the public and to build appropriate understanding of nanotechnology as applied to other areas of technology, through civil organizations, schools and public facilities.

This chapter first examines how the public should be engaged in ethical quandaries in cooperation with government and industry. It then outlines efficient educational strategies for teaching ethics in formal and informal education settings with the aid of existing resources, including civil organizations, school systems and public facilities. And, finally, it examines how resources and systems in other technologies and countries can be efficiently utilized.

PUBLIC ENGAGEMENT AND INITIATIVES

The humanitarian potential for nanotechnology is enormous, but so is the potential for its misuse. Mnyusiwall et al. (2003) raised ethical issues of nanotechnology at the very beginning of its development, stressing that ethical and social effects of nanotechnology have to be seriously explored. Likewise, the public should be involved in the consideration of the social implications of technology, since they bear directly on everyday life. There should be active communication between the public, government and industry, as well as other parties, so as to determine the best alternatives for yielding benefits and preventing the misuse of technology. Since the public comprises groups of beneficiaries from the developed technology, the public should actively engage the government and industry, demanding appropriate actions. In addition, journalists need to be involved in the early stages of the development and application of nanotechnology since they have an important influence on public perceptions about the benefits of such technology.

NGO activities

NGOs can play a vital role in reinforcing people's ethical views regarding technology and its effects on human lives and societies. Their actions can also provide regulations and guidelines to prevent the misuse of nanotechnology. NGOs should monitor potential consequences of rapidly developing S&T and call for environment- and human-friendly science by taking part in policy-making, implementation and evaluation in S&T, fields often dominated by the government, industry and the S&T elite. NGOs have to develop models to activate civil participation in S&T policy-making. They enable the public to obtain reliable and understandable information about technological innovations.

NGOs participate both directly and indirectly in decision-making about technological developments and implementation, acting as civil watchdogs. They have to activate the civic participation in the social dialogue on the enactment of laws to reduce the risks inherent in present and future nanotechnology. As of now, public awareness of biotechnology is relatively high, but few would recognize the concept of nanotechnology. To enhance citizens' awareness of ethical issues related to nanotechnology through public initiatives and involvement, NGO activities are vital. The following are examples of their activities in some countries.

In the Republic of Korea, diverse NGOs have actively played a key role in bringing together scientists, science-sociologists, science educators and students majoring in the natural sciences. They have organized regular events to help people to understand social issues related to S&T. The People's Science Center has done so, for instance, proposing to discuss the government's legal actions in consensus conferences, workshops, press reports, broadcasts, forums and position statements with the participation of experts and members of the public. They issue weekly newsletters to inform the people of the updated agenda and actions, publish alternative educational materials for science teachers, and hold public lectures. These activities call for timely action by other developed countries in response to concerns about nanotechnology development (CDST, 2005).

A Teacher's Human Rights workshop entitled 'Civil Rights Education and Science Subject with STS Approach' was held by the Korean National Commission for UNESCO at the Asia-Pacific Centre of Education for International Understanding (APCEIU) in 2001. In this workshop, the Science-Technology-Society (STS) approach was used to develop a human rights model. The second aim was to develop critical thinking on the ethical issues of biotechnology, ICT and energy technology. The third purpose was to discuss issues emerging from current developments in S&T for education. The final aim was to stimulate networking among teachers to share experiences related to human rights issues. The workshop was attended by approximately fifty science teachers, government officials from local districts and representatives from various organizations, including the Korean Teachers & Education Workers' Union, NGO Science, People's Solidarity for Participatory Democracy (PSPD), STS, and Teachers for Environment. On the last day, teachers discussed productive strategies for human rights education. All participants were satisfied with the content of the workshop and wished to participate in similar meetings in the future (CDST, 2005).

In Canada, the Action Group on Erosion, Technology and Concentration (ETC Group) is one of the active NGOs which has advocated following a precautionary principle, namely that nano-element production should be discontinued until its safety is proven. The ETC Group has published *The Big Down*, a report that analyses nanotechnology development, industry and related social implications (ETC Group, 2003).

Government involvement in nano-talk

Recognizing the ethical and social implications of technological development, some governments have attempted to facilitate 'nano-talk' among the stakeholder groups in order to communicate and exchange views from different standpoints. Governments' determination to invest heavily in nanotechnology makes it imperative that all consequences of R&D, particularly their social implications, be considered.

In the field of nano-biotechnology, nanotechnology can be beneficial for prevention, diagnosis and treatment of illnesses and disabilities. It is good news that the development of nanotechnology enables early diagnosis and better treatment with transplants and cellular repair systems. Some avenues of research in nanotechnology include the incorporation of artificial materials or machines in human systems, such as the implantation of computer chips. Still, many people are sceptical about the modification of living systems and the prospect of implanting artificial materials or machines in humans.

To minimize the related ethical problems, however, governments should intervene and regulate experimentation with fabrication of and usage of brain implants. The implications of such implants go beyond usual safety and ethical issues, calling in addition for legislative alternatives for crimes committed under the influence of such implants. Vogel (2001) asks who will be held responsible if a person commits a crime under the influence of externally controllable brain implants. This implies that there should be a comprehensive social and legal outlook on potential side effects of the utilisation of this technology.

In some countries, such as the US, Australia and the UK, government-affiliated organizations have initiated educational activities to talk to the public and communicate their views and opinions. In 2000, former US President Clinton announced that he would support R&D in nanotechnology. In November of that year, national nano-development plans were approved and five main research areas identified, one of which concerned the social implications of nanotechnology. In this context, a Subcommittee on Nanoscale, Science, Engineering, and Technology under the National Science Technology Commission (NSTC) held workshops on the social implications of nanoscience and nanotechnology. In these workshops, issues of how nanotechnology would affect medicine, the environment, employment, national security, ethics, law and culture were discussed and educational materials were distributed (Roco and Bainbridge, 2001). Thereafter, NSF held a workshop on 'Nanotechnology: Revolutionary Opportunities and Societal Implications', at which participants continued the discussion of the aforementioned topics, and also addressed topics including public attitudes towards nanotechnology, methods of communication and educational strategies (Roco and Tomellini, 2002).

In the UK the Royal Society and the Royal Academy of Engineering have collaborated in workshops and conferences on nanotechnology and published literature with the support of the Department of Commerce and Trade in 2003. As a result, the report *Nanotechnology and Nanoscience: Opportunities and Ambiguity* was published in July 2004. It defines the boundaries of nanoscience and nanotechnology, describes the state of scientific knowledge on nanotechnology, identifies its potential applications, and discusses present and future environmental and control issues (Royal Society & Royal Academy of Engineering, 2004).

Ethical and social issues regarding health were explored as well. It is noteworthy that the activities were carried out independently of any governmental intervention, despite the fact that the government funded it. Also in terms of the content, the activities highlighted the negative aspects of nanotechnology as well as the ambiguity of the nano-future, and formulated precautionary principles for the control of nanotechnology (ETC Group, 2004).

Regarding nanotechnology policy-making, people participated 'from the bottom up'. Their call for government action has led the government to acknowledge the ambiguous side effects of nanotechnology and nan-materials, such as nanoparticles and nanotubes, and to announce plans to form a NIDG (Nanotechnologies Issues Dialogue Group) in the near future.

On 25 May 2005, in Yorkshire, UK, a group of nano jurors was set up. The Nano Jury is one of the systems to facilitate people's participation and expression. Twenty people were randomly selected and asked to participate in a five-week programme of lectures on nanotechnology use and risks. Thereafter, they were asked to make statements on the final recommendations regarding the proposed agenda. This kind of opportunity not only enables communication of nanotechnology-related issues to lay people but also allows exchange of different views and perspectives among interest groups. In the UK, many projects with a similar 'bottom-top' participatory approach are in progress (Royal Society & Royal Academy of Engineering, 2004).

Nano-dialogue for interest groups

At the moment, ethical issues are much less discussed than the development of S&T. This leads to a lack of dialogue regarding such issues and their solutions. Dialogue between research institutes, granting bodies and the public on ethical issues is necessary; all aspects of development and related results need to be explored from various professional and lay viewpoints. The aim of this ethical dialogue will be to stimulate people's critical thinking and consideration, and to increase their knowledge of potential harms and benefits of nanotechnology.

The interest groups will include scientists, researchers, physicians, scholars and industrial groups. The dialogue should produce scientific, economic, educational and industrial interactions. The intent should be to work in a collaborative effort to meet the common goals of all concerned. The need exists for an innovative mechanism to raise professionals' and industries' awareness of ethical consequences. For example, to provide medical care in an ethical and humane way, physicians need to be better educated about the ethical aspects of medical practice. Routine bioethics education for medical students and resident physicians, and continuing medical education for practicing doctors, should be in place to meet the identified needs.

In the UK the Office of Science and Technology (OST) has been carrying out various S&T-related projects focused on society and human beings, in which the government, scientists and the public have participated. In September 2004, the Sciencewise Grant Scheme was established to educate the public about ethical and environmental issues of technology. Among the NGOs, the Nanotech Engagement Group (NEG) has worked in collaboration with the NanoScience Center of Cambridge University, Policy Studies Institute at East Englia University for more than two years to pursue 'nano dialogues' (ETC Group, 2004).

In Australia the Department of Industry, Science and Resources has held a workshop on nanotechnology in Australian industry. Workshop participants discussed not only opportunities for the practical transfer of nanotechnologies based on collaboration and communication with other fields and groups but also the ethics of changes which may come about. Although no solution to the issues discussed was found, there was agreement that appropriate national, interdisciplinary actions to deal with ethical implications are needed. For it is important that there be communication between professionals or fields to ensure that the benefits derived from a technology can be maximized and the risks minimized.

ETHICS EDUCATION FOR NANOTECHNOLOGY

The United Nations has already taken steps towards the teaching of ethics. At the World Summit on Sustainable Development in Johannesburg in 2002, world leaders reaffirmed the need for sustainable education. An initiative to strengthen the teaching of ethics was approved at the 1999 World Conference on Science and the Use of Scientific Knowledge organized by UNESCO and the International

Council for Science. The statement clearly expressed that the ethics and responsibility of science should be an integral part of the education and training of all scientists (COMEST, 2003).

In most disciplines, education has proceeded by establishing a foundation and then building pyramids of knowledge step by step. It has promoted enhanced departmentalization in academia and thereby allowed each field to imprint its own way of thinking on its scholars. While attitudes towards ethics may differ according to one's cultural background and belief system, it is important to determine one's own position on ethical issues and then gradually learn to consider the social effects of the technology which surrounds us in daily life.

Developing a sense of ethics and individual standpoints should start as early as possible. What is presented to students and future scientists has to reflect a balance in value and meaning based on their own positions. To achieve this, educators, counsellors and administrators must first be educated concerning social and ethical implications so as to be able to formulate effective teaching strategies in the field of ethics education. They have to be able to present clear and thorough pictures of what S&T can bring to human lives in both positive and negative terms. They have to provide their students with training in critical perspectives and logical reasoning, problem-solving skills and the skills needed to draw an ideal boundary between what technology offers that is desirable and what is not. Early and repeated exposure to ethical concepts and their relevance for life would help students to develop insights and to decide what is appropriate regarding human beings' relationship to technology.

The purpose of teaching ethics

According to the World Commission on the Ethics of Scientific Knowledge and Technology (COMEST), the objective of educational programmes is clear: teaching ethics so as to develop students' ability to recognize and analyse ethical issues and reach sound ethical decisions. A programme has to increase students' awareness of ethical issues and enhance their moral sensitivity. Here 'students' does not mean only those in formal education, but also in informal educational settings. It is important that students become aware of normative dimensions so that they are able to identify both technical and ethical aspects, for example, of nanotechnology. With well-structured programmes, the students will learn to reflect critically about the most salient ethical issues in today's existence, particularly those that arise in the context of the countries in transition.

As mentioned earlier, ethical perspectives are different depending on the cultural and traditional background. Also there may be regional differences regarding what the actual ethical problems are felt to be. Yet such differences should be addressed through education (COMEST, 2003).

Formal education for nano-ethics

Ethics is a field of study, one that has never been more intensive than today (COMEST, 2003). Students should become familiar with different levels of ethics education and all the various notions and distinctions that one must take into account to reach sound ethical decisions, from basic subjects to argumentation and ultimate justification.

The middle and high school setting

The goal of today's science education is to enhance understanding of the nature and interactions within STS. From this new perspective on science, people can see that science is no longer a subject separate from society but plays a vital role within the socio-cultural context. Since the 1980s, STS has been has been an internationally recognized subject area, and efficient STS education materials and programmes have been developed (Roy and Waks, 1985; Vazquez-Alonso and Manassero-Mas, 1999).

In the Republic of Korea, STS was introduced with the 6th National Curriculum in 1992 at the elementary and secondary school levels. STS has been given even greater emphasis in the 7th Curriculum, which stated that STS could contribute to achieving four major educational objectives of the national basic curriculum. With this idea, 'Living and Science' in the high school curriculum was based on STS perspectives (Choi, 1996).

The introduction of newly developed teaching and learning materials on the ethics of technology has proved to be an efficient means of establishing ethics education in the academic setting. Such educational initiatives regarding biotechnology have already been carried out. For example, the European Commission (EC) has been actively contributing by publishing European Initiatives for Biotechnology Education (EIBE) units, learning materials used by high school students (aged 16–19). EIBE units cover experimental activities, recent science research, games and role playing. In addition to these materials, the EC has also provided science information, news articles, related literature reviews, ethical guidelines, students' activity plans, for immediate introduction into the classroom (EIBE, 2002).

In the US, the Biological Sciences Curriculum Study (BSCS) has received a grant from the National Human Genome Research Institute (NHGRI) and the Department of Energy and has been developing a high school curriculum for ELSI on biotechnology since 1992. So far, ten different educational programmes have been developed and distributed among over 50,000 teachers for free. Likewise, the Applied Professional Ethics Center at the University of Tennessee has developed 'The Social Impacts of Human Genetics Engineering' programme, which contains various visual aids and materials for use in discussion classes in high schools (NHGRI, 2001).

In the UK the National Center for Biotechnology Education (NCBE) was established in 1985 to facilitate and enhance people's science literacy related to biotechnology. Since then, the NCBE has developed experimental equipment and learning materials to be used in the school setting; they have been sold to twenty-six countries so far. According to a study by the University of Leeds, the NCBE has received high marks for introducing innovative learning programmes for school teachers (NCBE, 2002).

The Kennedy Institute of Ethics at Georgetown University has developed the High School Bioethics Curriculum Project (HSBCP) for use in high school curriculum, mainly regarding biotechnology-related ethical topics. This programme is composed of four parts (research ethics, eugenics, genetics-related disease and organ transplantation) and between five and eight cases. Teachers were asked to prepare the content, identify each course's objective, develop discussion topics and compose reading list (Downie, 1993).

Another example is the Department of Genetics at the University of Washington in Seattle, which provides education programmes for middle school, high school, and the general public. Among them, the High School Human Genome Project focuses on DNA analysis, biotechnology-related ethics, such issues connected with Huntington's disease, and human duplication.

These kinds of educational strategies have stimulated further educational action to address ethics-related subjects in high school curriculum. In Japan, *Guidelines to Bioethics* was published as high school teaching and learning materials and were distributed to 500 schools. Since then, the need to offer bioethics classes in high schools was seriously considered (Mays, 1998).

The foregoing examples are of educational strategies concerning various technology areas, mostly biotechnology, which were formulated and implemented in the early 1990s with relevant ELSI issues. Thus far, no teaching or learning materials on nanotechnology have been developed for use in middle or high school curricula.

Education at the university level

Elective courses: Students need to gain science-related knowledge about contemporary life, and this need should be met through suitable educational programmes not only in primary and secondary education but also in higher or tertiary education. Such programmes would further enhance students' interest in STS and also facilitate research. Through these interventions, educational institutions will be able to meet the demands of society, and further develop materials for comprehensive education in STS.

At the University of Beijing, graduate and doctoral students in the Center for Scientific Law Research have been obliged to take a course on biological ethics since 1991. The course is intended to improve students' scientific and human experience.

In the Republic of Korea, material for an elective course was developed with the aim of improving students' understanding of STS in the socio-cultural context and to focus on the nature of science; emphasis was placed on science as it is encountered in daily life rather than scientific theory (Choi et al., 2005). In the study, the STS course was provided to 265 freshmen and sophomores majoring in the humanities, the fine arts, the social science, the natural sciences, or engineering. The STS course significantly increased students' science literacy. Following the course, students were asked whether they considered such classes on STS necessary. Although analyses of the responses according to student major showed that differences in their perception of the need for STS education depended on their majors, academic backgrounds, levels of interest in and knowledge of science, 97.7 percent of the participants nevertheless responded that there is a need for STS education at the university level (Choi et al., 2005).

Currently, science-related classes offered at the university level are unable to enhance STS literacy. To redress this situation, continuous

STS education from elementary school up to the college level is needed. Additionally, classes (elective or obligatory) focused specifically on STS should be developed and provided at the university level. First, the classes should be taught by ethics experts. Although ethics is of concern to all, not everyone is qualified to teach ethics. Ethics teachers should have a broad competency in ethical theories and ethical argumentation. Second, funding is necessary for the purchase of ethics-related resources and materials to be used in class. Instead of being a lecture course, the class should give students be taught with visual aids and through direct, hands-on experience with the topics under discussion.

Degree programme in ethics: Since September 2003, the University of Utrecht, in the Netherlands, has been offering a Master's degree programme in applied ethics. The year-long programme, which is conducted in English, is aimed at students from various backgrounds, including medicine, veterinary medicine, biology, philosophy, theology and law. It provides in-depth knowledge of different ethical theories, as well as important methods and discussions in applied ethics.

At the University of Zagreb, in the Republic of Croatia, a Master's degree programme in health, human rights and ethics is offered to professionals from South-East Europe. It is a two-year programme focusing on ethics in relation to healthcare and human rights. It consists of twelve modules on the state of the art in bioethics, and five modules on practical skills. The programme aims to foster the critical reflection and moral sensitivity of students.

Unlike elective courses, these Master's programmes educate professionals in ethics. Most existing programmes require the prospective students to have a Bachelor's degree in a related field. If as undergraduates students have studied other disciplines and taken ethics-related classes, it is easier to provide them with a more structured sequence in their graduate-level studies in ethics.

Research ethics within the school curriculum

Any effort to reinforce ethical, legal and social issues in research has to start with students as they enter higher education. In that context, students are able to sharpen their analytical skills and to develop and exchange knowledge and views with other undergraduates, as well as graduates, postdoctoral fellows, junior faculty and senior investigators. This can be further stimulated by presenting career awards and training.

With the increase in research grants and government funds for research, emphasis on ethical, legal and social issues (ELSI) in research should also increase.

In graduate or doctoral programmes, thesis or dissertation research is mandatory to obtain a degree. Future scientists and researchers should be exposed to ELSI aspects of research in technology in the course of their own research. As COMEST (2003) has proposed, students should consider the ethical issues raised in their thesis or dissertation, both the internal problems of research ethics and the external problems that arise in connection with likely applications of the results they describe in their work.

Researchers need to assign a certain percentage of their budgets to solving consequential ethical issues. For example, the Human Genome Project benefited from the decision to allocate 3–5 percent of research funding to ELSI research. In the case of Asia, despite the fact that a certain amount of government funding is allocated every year for nanotechnology development, no percentage of nanotechnology budgets are allocated for the consideration of ELSI; therefore, steps should be taken towards a serious consideration of this issue.

In Bloomington, Indiana, the Association for Practical and Professional Ethics has published six volumes of case studies in research ethics. Each summer since 1996, the association has convened a workshop in Bloomington for graduate students in the sciences and engineering. Each participant writes a case study and commentary. Individual cases focus on a variety of topics, including issues in research using human subjects and intellectual property (COMEST, 2003).

Public education through university-affiliated programmes

Professionals, scientists and experts in technology should be involved in continuing education regarding ELSI. Some countries have already initiated this action for some types of technology, though not for nanotechnology. University-associated bodies and institutes can also provide programmes to the public. For example, the Genetics Science Learning Center (GSLC) is located in the Human Genome Research Genesis Center (HGSRGC) at the University of Utah. HGRGC provides both professional and public lectures and fora on bioethics and research guidelines. University programmes of this kind provide comprehensive content developed by researchers and experts in the university. Since it is important to study the social and ethical implications of technology from different perspectives, these programmes should be developed using interdisciplinary approaches.

Continuing education or re-education for ethics competency

Besides formal learning in degree programmes, it is also necessary to provide relevant ethics-related classes to professionals and researchers in the context of continuing education programmes. Again, one can find model examples in the field of biotechnology. One relevant example is the GenEd (Genome Education) programme developed not only to educate ELSI experts in a formal, university setting but also to educate medical health professionals in the context of less formal, continuing education programmes (NCHPEG, 2002). The GenEd programme also evaluates the appropriateness of genome education and training systems for professionals who did not major in genetics in the medical health fields. In the EU, professional training has been mandatory for medical experts in England and the Netherlands, while it has been optional in Sweden, France and Germany (Micklos and Carlson, 2000).

Continuing education programmes aim to bring professionals up to date about recent outcomes in ELSI in S&T. Such programmes help them to gain a better understanding of the possible consequences of S&T. And thus it is imperative that professionals and scientists involved in nanotechnology be obliged to participate in such continuing education.

COMEST (2003) has suggested that teaching ethics for scientists requires not only a solid competence in ethics but also a thorough knowledge of the science with respect to which ethics is being discussed. Dual competence in science and ethics is important in order to be able to deal with ethical issues in a meaningful and effective manner. One has to know ethics well enough to be aware of crucial distinctions and considerations that make the difference between good and bad arguments. Without such dual competence, scientists will tend to think of ethics as a matter of expressing one's convictions and ethicists will tend to present arguments and considerations that have little bearing on the real issues. Through re-education or continuing education opportunities, scientists and researchers will deepen their knowledge of ethics and so move closer to achieving competency in both areas (COMEST, 2003). On the other hand, whatever form the education may take, much can be achieved by co-teaching, where scientists and ethicists teach together. Likewise, much can be gained within scientific research by having ethicists work as members of research teams (Roy and Waks, 1985).

Informal education for nano-ethics

Informal learning in science takes place in a variety of contexts outside formal institutions and has become quite significant as a complement to formal curriculum. The value of informal learning in promoting science education and science is increasingly thought comparable to that of formal education (Wellington, 1994).

There is a great variety of sources for obtaining educational experience of science outside school curricula. One effective source of science education can be the media, such as television programmes, radio, newspapers, film and the Internet. Science museums or science centres, natural history museums, planetariums, zoos and aquariums, botanical gardens and parks, nature centres and environmental education centres, and scientific research laboratories – all can provide practical learning experiences. Community-based organizations and projects, including youth organizations and community outreach services, can also be places for science learning and education. In a relatively short time, an impressive number of science centres have emerged in every country, each with its own distinctive character and emphasis in science (Song et al., 2002). As social education and continuing education have gained in importance, education has also become a central part of the museum's function.

The educational role of science centres and museums

Early exposure to the concept of science and to various methods of critical thinking and decision-making can start gradually, depending on the age level. As the museum's role has expanded from a simple display of objects, it has developed its own significance at the social, economical and political levels (Song et al., 2002). The role of science museums is now changing from being an exposition venue to becoming an active and experiential educational setting for informal learning. This may be the reason why science centres have been growing rapidly in recent years. At the social or cultural level, they help people to understand evolving technologies and to predict future consequences of future advances as well as to consider them in view of human development, dignity, rights and other ethical issues. In this respect, a science museum is a place where people can obtain objective information on various facets of scientific development and further develop insights into ELSI. As for nanotechnology, science museums would do well to design their exhibits in view of the ethical and social implications of nanotechnology so as to stimulate students and citizens to engage in balanced thinking about the costs and benefits of the research agenda.

Despite the fact that museum exhibitions are often sponsored by industries and commercial institutions and thus may harbour a bias in favour of the sponsors, museums have shown themselves to be extremely efficient and comprehensive in introducing science to the public. The advancement and rapid proliferation of S&T in almost every aspect of human activity has given rise to a host of associated ethical and moral issues.

Many science museums or centres in the US, Germany and the Republic of Korea deal with various aspects of nanotechnology. They have effectively introduced the concept of nanotechnology, though they have not dealt sufficiently with its ethical implications. Some critics have suggested that for many science centres, presentations and cases convey the benefits of nanotechnology in order to attract people's attention and interest, rather than raise potential ethical or social issues. However, because ethical issues concerning S&T have become more urgent in contemporary society due to the rapid development of S&T, people need to understand science in order to deal with the many complex science-related issues that confront citizens of modern democracies. At the Museum of Science in Boston, there is an introductory exhibit on science and the carbohydrate nanotube. The Deutsches Museum (or German science museum) in Munich has a nano experiential exhibit in which atomic technology resulting from nanotechnology and genetic research are presented. The museum provides information about innovations and marvellous phenomena that may arise when nanotechnology is combined with medicine, new materials and IT. At the National Science Museum of the Republic of Korea, there are active experimental exhibits to help visitors to gain a deeper understanding of nanotechnology; to this end there are also visuals, panels (such as 'Spider-web and nano', 'What is my height in nanometres?'), and case presentations of nanotechnology applied to biotechnology. These experiments introduced people to the unlimited potential of the nanoworld, while also showing some new products developed with nanotechnology, such as the micro-robot. Certainly, these exhibitions have helped people to understand that nanotechnology is a rapidly advancing field that is affecting all of human life.

An examination of four world-renowned science museums, analysing by type and form of presentation the educational strategies for ethics education incorporated in the offered programmes, showed that the museums did present ethical issues related to the development of S&T, though mainly in the fields of biotechnology and environmental issues (Choi, 2004).

Table 7.1. Presented ethical themes and topics in the museums

	BT	ET	Medicine	Energy	Health	NT	Astronautics	Food	Agriculture	DT
A	✗	✗	✗	✓	✗	✓	✓	✓	✗	✗
B	✗	✗	✗	✗	✓	✗	✗	✓	✓	✓
C	✗	✗	✗	✓	✗	✗	✓	✓	✓	✓
D	✗	✗	✗	✗	✓	✓	✗	✗	✓	✓

A: The Science Museum, London
B: La Cité des Sciences et de l'Industrie, Paris
C: The Museum of Science, Boston
D: Deutsches Museum, Munich
✗ – absent
✓ - present

Other types of technology, such as nanotechnology, were discussed only in a rudimentary way and without dealing with ethical concerns related to them. Hence, modern science museums should play an active role not only in educating the public about contemporary science for its own sake but about also in stimulating the formation of ethical perspectives and ethical debate on the development of S&T.

Formulation of programme content

Types of programmes with ethical themes: Various events have taken and continue to take place in museums concerning themes and topics under discussion here, events that have been staged to communicate those themes and topics in the most efficient way. Demonstrations and on-line exhibitions, panels and simulations were held most frequently to deal with ethical issues. Museum programmes have also included lectures, plays or shows, debates and workshops. Museums should actively search for and develop the most efficient ways to deepen the public's understanding of science-related ethical issues. At the four science museums just discussed, the most frequent type of presentation dealing with ethical issues were exhibitions, panels and simulations, followed by demonstrations and lectures. All of those museums had the characteristics necessary to play an active role in educating the public about ethical issues in science (Choi, 2004).

Table 7.2. Types of presentation for ethical issues on S&T

	Panel	On-line Exhibition	Demonstration (exhibits)	Lecture	Play	Debate	Workshop	Simulation
A	✗	✗	✗	✓	✗	✗	✓	✗
B	✗	✗	✓	✗	✓	✗	✗	✗
C	✗	✗	✗	✗	✗	✓	✓	✗
D	✗	✗	✗	✗	✓	✓	✗	✗

A: The Science Museum, London
B: La Cité des Sciences et de l'Industrie, Paris
C: The Museum of Science, Boston
D: Deutsches Museum, Munich
✗ – absent
✓ – present

Age-appropriateness: The age range of visitors to museums is very wide. Since science museums are open to the general public, they have to design programmes with the broadest appeal, thus matching all age levels, in order to make the experience enjoyable and challenging for as many people as possible. For children, the activities should be experiential so that they can learn through hands-on activity and manipulation. The content should be enjoyable, challenging and thought-provoking. Selecting the most efficient medium for different age groups will make ethical exploration most efficient. For example, plays or shows would be more appropriate to communicate ethical issues to children, whereas lectures and panels would be most appropriate for adults.

Open questions for self-inquiry: S&T are still developing and will continue to do so. Ethical issues involve not only the identified present problems but also questions about what may happen. Rather than finding the 'right' answer to the ethical issues, it is important to explore them and ascertain different answers where appropriate. Therefore, exposure to stimulating questions and the exploration of creative solutions are significant in ethics education. This also means that objective perspectives on certain issues, including both positive and negative outcomes, need to be incorporated in museum programmes. In recent years, there have been a number of proposals to extend the role of museums in society and assign them an active role in educating the public on scientific literacy (Henriksen and Frøland, 2000).

Online education

With the increasing number of Internet users, online education on ELSI of nanotechnology can be an effective strategy. Considering the power of the Internet these days, the effect of online education can be amazing when utilized efficiently with appropriate content. In fact, this approach has been taken with other types of technology, especially biotechnology. At the University of Kansas, the medical centre has a Genetics Education Center (GEC) which provides various educational activities. GEC has organized the Human Genome Teacher Networking Project (HGTNP) with the grant from NHGRI. HGTNP has been focusing on the ELSI of the HGP and provides relevant online information to students, teachers and the public.

Another example of the use of online education regarding ethical aspects of technology is the Cold Spring Harbor Laboratory's learning centre called Dolan DNA Learning Centre. The Dolan centre is the first place in the world to provide genetics education to students, teachers and the public in formal and informal settings. Since 1998, all its eugenics research material has been digitalized and provided to the public. This service has contributed greatly to the understanding of the importance of human privacy (Micklos and Carlson, 2000).

Online education will certainly assist ethics education in developing countries. Many countries, especially the developing ones, would need external support to develop programmes to build up professionals' competence in ethics. While there is a lack of qualified teachers and dynamic materials for ethics programmes, and ethics researchers and teachers often have little opportunity to participate in international events to expand their knowledge, involving them in online education will certainly help to overcome these obstacles. Since the Internet is part of most people's everyday life, organizing meetings in cyberspace would likely be a practical and efficient strategy.

CONCLUSION

While promising massive changes in our lives, technology involves ambiguous consequences and engenders suspicion and fear. Nanotechnology offers enormous potential for positive change, especially in the areas of healthcare and medicine, but open public discussion is needed regarding the benefits and risks of this new technology to ensure public understanding of its development. The ELSI of nanotechnology involve privacy, security, environmental and metaphysical questions, all which require detailed analysis and discussion and ultimately careful and rigorous regulation.

In this chapter, some strategies for public action and for education were explored. Existing public resources, such as civil organizations and science-related facilities, can play active roles in ELSI education. NGOs can actively bring the government, industry, scientists, researchers and other interest groups together for nano-dialogues, and educate the public through various NGO activities, including statements, presentations, lectures, seminars, workshops, exhibitions and consensus conferences.

As for ethics education, it can occur in formal and informal settings. Formally, it can involve teaching ethics from middle school onward, and offering both elective and major courses in ethics at the university level. Informally, science centres or museums can play a significant role in public education, presenting various ethical issues of technology to visitors of all ages. The content should be age-appropriate, experiential and formulated in the way that best suits the theme in question, such as hands-on activities, exhibitions and role-playing. Ethical problems have become matters of course in modern society, since S&T are developing rapidly. The need for education on ethical issues is becoming more and more urgent. Citizens of modern societies need to understand science in order to deal effectively with the many complex science-related ethical issues (Micklos and Carlson, 2000).

BIBLIOGRAPHY

Center for Democracy in Science and Technology (CDST). 2005. *Newsletter*, Nos. 47–54, May-June. http://cdst.jinbo.net. (in Korean).

Choi, K. 1996. *STS Education: Understanding and Application.* Seoul, Kyohaksa.

———. 2004. Developing active role of science museum for educating ethical issues on science and technology. *Journal of Korean Association for Research in Science Education*, Vol. 25, No. 3, pp. 336–45 (in Korean).

Choi, K., Chong, J., Chang, H. and Lee, K. 2005. Need and significance of STS education at the university level. *Journal of Korean Association for Research in Science Education*, Vol. 25, No. 6, 650–57 (in Korean).

COMEST (World Commission on the Ethics of Scientific Knowledge and Technology). 2003. Teaching of Ethics: Second meeting of the COMEST Working Group. Minutes of the second meeting of the COMEST Working Group on the Teaching of Ethics, Oslo, 10–12 May 2003. Paris, UNESCO. http://portal.unesco.org/shs/en/file_download.php/84cba1f1a29d9d7f68f0a64693e7471eCOMESTEthicsTeachingWG2.pdf (Accessed 20 December 2006.)

Downie, J. R. 1993. Teaching of bioethics in the higher education of biologists. *Journal of Biological Education*, Vol. 27, pp. 34–38.

ETC Group 2003. *The Big Down: From Genomes to Atoms*. Ottawa, ETC Group. See http://www.etcgroup.org/upload/publication/104/01/littlebigdown.pdf (Accessed 20 December 2006.)

———. 2004. UK report: More hits than misses on nanotech. News release, 29 July. http://www.etcgroup.org/upload/publication/89/01/nr_rsfinreport.finaldoc.pdf (Accessed 20 December 2006.)

European Initiative for Biotechnology Education (EIBE). 2002. http://www.eibe.info (Accessed 25 January 2006.)

Henriksen, E. K. and Frøland, M. 2000. The contribution of museums to scientific literacy: Views from audience and museum professionals. *Public Understanding of Science*, Vol. 9, No. 4, pp. 393–415.

Mays, M. E. (ed.). 1998. *The Genetics Revolution: Programs and Issues for the Community College*. Washington, DC, Community College Press.

Micklos, D. and Carlson, E. 2000. Engineering American society: The lesson of eugenics. *Nature Reviews*, Vol. 1, pp. 153–58. http://www.dnalc.org/downloads/naturearticle.pdf (Accessed 20 December 2006.)

Mnyusiwalla, A., Daar, A, S., and Singer, P. A. 2003. 'Mind the gap': Science and ethics in nanotechnology. *Nanotechnology*, Vol. 14, pp. R9–R13. http://www.utoronto.ca/jcb/home/documents/nanotechnology.pdf (Accessed 20 December 2006.)

National Centre for Biotechnology Education (NCBE). 2002. The history of the NCBE. http://www.ncbe.reading.ac.uk/NCBE/NCBE/history.html (Accessed 20 December 2006.)

National Coalition for Health Professional Education in Genetics (NCHPEG). 2002. The Future is Now. Fifth Annual NCGPEG/GROW Meeting, Bethesda, Maryland, USA.

National Human Genome Research Institute (NHGRI). 2001. Ethical, Legal, and Social Implications (ELSI) Research Program. http://www.genome.gov/10001618 (Accessed 20 December 2006.)

Roco, M. and Tomellini, R. (eds.). 2002. *Nanotechnology: Revolutionary Opportunities and Societal Implications*. Report of the Third Joint EC-NSF Workshop on Nanotechnology, in Lecce, Italy, 31 January–1 February 2002. Luxembourg, European Communities. ftp://ftp.cordis.lu/pub/nanotechnology/docs/nano_lecce_proceedings_05062002.pdf (Accessed 20 December 2006.)

Roco, M. C. 2003. Government nanotechnology funding: An international outlook. 30 June. Washington, DC, National Science Foundation. http://www.nano.gov/html/res/IntlFundingRoco.htm (Accessed 20 December 2006.)

Roco, M. C. and Bainbridge, W. S. (eds.). 2001. *Societal Implications of Nanoscience and Nanotechnology*. Proceedings of a workshop organised by the National Science Foundation, 28–29 September 2000. Dordrecht, Kluwer Academic Publishers.

Roy, R. and Waks, L. J. 1985. The A. B. C.'s of science, technology, society. *FORUM*, Vol. 13, No. 4, pp. 1–4.

Song, J., Oh, W., Cho, S. and Koo, S. A. 2002. *Survey and Developing DB System for the Facilities Supporting Students' Out-of-school Science Activities in Korea*. Seoul, Korea Science Foundation.

The Royal Society & The Royal Academy of Engineering. 2004. *Nanoscience and Nanotechnologies: Opportunities and Uncertainties*. 29 July. London, The Royal Society & The Royal Academy of Engineering. http://www.nanotec.org.uk/finalReport.htm (Accessed 22 December 2006.)

Vazquez-Alonso, A. and Manassero-Mas, M.-A. 1999. Response and scoring models for the 'Views on Science-Technology-Society' instrument. *International Journal of Science Education*, Vol. 21, No. 3, pp. 231–47.

Vogel, V. 2001. Social impacts of nanotechnology in education and medicine. M. C. Roco and W. S. Bainbridge (eds.), *Societal Implications of Nanoscience and Nanotechnology*. Dordrecht, Kluwer Academic Publishers, pp. 143–47.

Wellington, J. 1994. Using informal learning to enrich science education. J. Wellington, J. Henderson, V. Lally, J. Scaife, S. Knutton and M. Nott (eds.), *Secondary Science: Contemporary Issues and Practical Approaches*. London, Routledge, pp. 284–94.

Chapter 8
EARLY ASSESSMENT AND POLICY-MAKING

Michèle S. Jean, Édith Deleury and Diane Duquet

Is it possible for ethics to proceed in tandem with the evolution of a new technology? Is it possible, before new products are even defined, to keep pace with scientists in determining what these potential products might be, as well as the questions to be raised for their ethical evaluation? This is the challenge the Quebec Commission on Ethics in Science and Technology (CEST) of the Government of Quebec in Canada (see Annex 1) set itself in 2004, when it decided to prepare a position statement on the ethical issues of nanotechnology.

To evaluate the ethics issues involved in a new technology before it has been widely disseminated has important advantages. For example, early intervention permits the implementation of the necessary frameworks for the responsible development of the technology by its promoters (public or private); eventually, it can ensure greater acceptance of it on the part of the general public, because the dialogue between the various concerned parties had been underway from the beginning. Yet it is rare that this is the case, which is the source of the criticism that ethics always lags behind science.

PREPARING THE FIELD FOR A VALUABLE ASSESSMENT

It is well known that different views exist on how the ethics of nanotechnology should be assessed. Despite the difficulties, CEST has, for the time being, adopted Preston's approach, namely:

> [There] are suggestions that nanotechnology is so radical and its disciplinary foundations so unusual that it requires an entirely new ethical framework,

> one tailor-made for the issues (Khushf, 2004). So electric is the buzz around nanotechnology that some of those cognisant of its implications want a completely clean ethical slate for their discussions. Here I argue a different case. (Preston, 2005)

CEST has decided that it is possible and useful to tackle the subject of nanotechnology so that ethics be involved in the debate early on, at least in the province of Quebec. This should then put the Commission in a position to engage the public in a well-informed debate that could avoid a simplistic rejection of the technology based on false premises.

CEST's decision to prepare a position statement on the ethical issues of nanotechnology was based on a number of factors which the Commission put into perspective and evaluated: the state's interest in S&T development in this area and its investments in it; the attention the media dedicate to the subject and the public's level of understanding (or misunderstanding) of the subject; the attention the subject is commanding at the international level (investment, reports, studies, consultations, etc.); the impact on society of the information assembled; and the beginnings of ethical inquiry. The importance and potential benefits to be considered also weighed in favour of examining the ethical issues of this new technology at such an early stage in its development.

Once it had taken its decision, CEST proceeded to identify experts in the field who, in concert with its members, could form a working group capable of realising the diverse facets of its mandate:

1. To understand the field (applications, actors involved).
2. To provide an overview of existing regulations and guidelines at national and international levels (risk management strategies, application of the precautionary principle, etc.).
3. To take stock of what is going on in the province of Quebec.
4. To determine risks and promises for human health and the environment, work organisation, governance, etc.
5. To assess the level of knowledge of the population in Quebec.
6. To identify the values and ethical issues at stake.
7. To propose recommendations.

To create the working group, CEST had to address an important characteristic of the field of nanoscience and nanotechnologies: its multidisciplinarity, which requires collaboration between biologists, physicists, chemists, material scientists and information technologists, among others. To the field's multidisciplinarity must be added the diversity of applications in, for example, the health, environmental, agricultural, food, energy and national security sectors, as well as the repercussions that must be considered in relation to work organization, consumer protection and governance, but also the future of humanity and the type of society which will be passed to future generations (see Annex 2). In short, it was necessary to create a working group that could understand the complexity of the subject not only from a scientific perspective but also from an ethical, social and even economic point of view. Given the inherent size limit (between eight and twelve members) for this type of committee so that it would be operational, the Commission supplements the knowledge and representation of members by holding roundtables or thematic presentations with specialists to address the different facets of the subject of study (see Annex 3).

The working group began its work in December 2004 and should complete it in April 2006. A total of twelve meetings, including six roundtables or presentations, will allow the group to produce a position statement and make recommendations, which will be submitted to CEST for discussion and adoption. The submission of the position statement to the Quebec government and its diffusion to the general public will take place in May 2006.

It can be argued that there is no need to look at these issues because international studies can be used to assess the field. This is not CEST's view. Each country, province or community has to go through the process and make it their own: the process must be grounded in its cultural, social and even economic context. Indeed, what is going on at the international level will be taken into account, but ultimately it will be the country or province that will decide how far it will go in research, investment, university programmes, etc. in the field under study.

ASSESSING THE ETHICAL AND SOCIAL IMPLICATIONS OF NANOTECHNOLOGY

To prepare a position statement – notably in the context of implementing a new technology – the ways to proceed for the evaluation of the ethical and social consequences are essentially the same, regardless of whether the technologies in question have already been implemented, are in the process of being implemented, or are just beginning to be developed. However, there may be important differences regarding the level of uncertainty about the nature of the effects (actual or potential) to be considered in a particular stage of development of a science or technology, and regarding the type of framework (i.e. law or regulation), both existing and desired.

CEST's approach to ethical evaluation

At CEST the ethical assessment process that has been adopted to combine both reflection and dialogue is the following: understanding the situation, considering the realizations and promises and ascertaining the risks (real or potential), identifying and analysing the consequences, pinpointing and clarifying the values at issue, characterizing value conflicts and hierarchies, and developing a practical rationale to justify decisions and recommendations (CEST, 2003, p. 57). This value-based approach is the foundation of the CEST's decision-making process and of the justification for its recommendations.

Its application to the ethical evaluation of nanotechnologies

CEST's approach to ethical evaluation was put into practice in the preparation of its position statement on nanotechnology. From the outset, the Commission considered the following elements relevant to the field:

- The fact that ethical reflection and evaluation must be situated prior to nanotechnological applications (upstream) and, at the same time, after certain applications (downstream).
- The specificity of nanotechnologies:
  – their size at the nanometre scale and therefore the lacunas of existing instruments to see and manipulate nanometric material;
  – new and unknown properties that materials demonstrate at this scale;
  – the multidisciplinarity (physics, chemistry, biology) of nanotechnology and the convergence of disciplines (notably biotechnology, information sciences and cognitive sciences) it brings about;
  – the methods of manufacture (notably self-assembly as occurs in nature);

– world trends, and the diversity of practices and frameworks currently in place, as well as the diversity of objectives being pursued.

Stages in exploring the subject

Understanding the situation (at the scientific, economic and normative levels)

The goal of this first stage is to fully grasp all facets of the issue under study and to set it in a global, well-documented context. It entails a literature review and consultations with experts in the working group, as well as external experts through roundtables on a variety of themes: health, environment, work organization, economics, governance, etc. At the end of this stage, CEST agreed to concentrate on four sub-areas of nanotechnology: nanomaterials, nanoelectronics, nanobiotechnologies and nanometrology.

This stage also allows for a review of the state of regulation which may guide technological development and assure adequate protection to the diverse actors and activity sectors concerned. International conventions, charters of rights and liberties, laws and regulations, good practice guidelines, etc., are collected. In the case of nanotechnologies, other instruments must also be considered, such as the precautionary principle and lifecycle assessment, which can be useful to guide technological developments.

Considering the realizations and promises, ascertaining the risks (real or potential)

This stage builds upon the literature review, the consultation of experts and the expertise of the members of the working group. The goal here is to determine the applications of nanotechnologies currently on the market, and also what applications could potentially flow from the technology in the near future. Further, it serves to determine the risks (real or potential) that may be associated with the applications of nanotechnology in its current state of development, as well as its foreseeable state of development, based on ongoing research.

At present, nanotechnologies are used mainly to improve certain characteristics of products, such as sunscreen, textiles, glass, rubber and electronics, or to endow them with new properties, such as resistance, lightness, malleability, non-inflammability, and many others. Health, the environment, work organization, transport and electronics are domains that could benefit from the innovations of nanotechnology. It would be irresponsible, however, to ignore the risks that can accompany the production, distribution or disposal of these new products, notably with respect to their potential effects on the environment or on consumer or worker health, or even on the transformation of the human being – with optimization of physical or intellectual performance that could eventually be possible with certain applications of nanobiotechnology, for example.

Identifying and analysing the consequences
This stage is marked by discussion on the subject and the documentary research (consultation of experts as needed) to corroborate or refute the apprehended consequences, actual or potential, as well as their incidence on society, present and future.

In the case of nanotechnology, these consequences concern, first, the toxicity of nanoproducts for health and the environment. However, the consequences of nanotechnology are also apprehended in work organization, economic development (marginalisation of the least well-off), governance, privacy protection, and the transformation of the human being, not to mention possible repercussions on developing countries (such as the enlargement of the technology gap with industrial countries).

Pinpointing and clarifying the values at issue
In determining a consequence flowing from a technological application, there is at least one but often numerous values to promote or protect; discussion allows for an analysis of what these values are and how much importance each member attaches to them both in principle and in the context of specific decision-making.

Numerous values are at issue in the context of nanotechnology: the protection of health and the environment, transparency, equity, respect for persons and their integrity, respect for privacy, the exercise of democracy and social responsibility, accessibility and many others, but also economic development and its impact on improving quality of life in the population. Will the development of nanotechnology and of some of its applications support these values or undermine them?

Characterizing value conflicts and hierarchies
The values collectively retained must confront one another with regard to the issue being studied to determine whether there may be a conflict

between two values at the time of formulating recommendations. For example, can the values of human health, the environment and economics all be promoted concurrently or does the choice of one of these values necessarily undermine the promotion or protection of another value in the context of a given recommendation? Is it possible to reconcile divergent values? On what basis does one reconcile a hierarchy of values?

In the preparation of its position statement on nanotechnology, CEST has now reached this stage, which accompanies the formulation of recommendations aimed at favouring ethical decision-making about technological development. This stage could be enriched from results obtained through public consultation, while taking into account the subjects of the study and the time and resources (human and financial) available. This will be discussed later on.

Developing a practical rationale to justify decisions and recommendations
Practical reasoning constitutes the essence of the approach, and affects both components of the decision: the choice of ends, and the choice of means. In ethical deliberation, the choice of means must take into account their effectiveness to achieve the desired goal, as well as the ethical character of those means. The end does not justify the choice of means, and it is the achievement of the best equilibrium between conflicting values that is sought in identifying the effectiveness of a mean.

In a domain such as nanotechnology, and because of the level of uncertainty regarding the effects some of nanotechnology's applications may potentially have on the human being, its environment, and society in general, it is not always possible to formulate recommendations that are concrete, realistic and attainable. As in the position statement that the CEST prepared on genetically modified organisms (GMOs), it is possible to formulate 'cautionary notes' on some questions that decision-makers must consider in order to make a formal decision that takes into account the ethical issues put forward.

CONSULTING THE PUBLIC

In a press release UNESCO (2003, p. 11) declares that

> Governments of Member States and legislators with whom decision making ultimately rests, have major responsibilities in this regard [bioethics]. In this process, they must see to it that citizens have an opportunity for informed, pluralistic public debate, and must take into account the various schools of thought, value systems, historical and cultural backgrounds, and

> philosophical and religious convictions that make up our various societies. Clearly, bioethics must be based on the practice of democracy and the active participation of all citizens.

Further, the *Universal Declaration on Bioethics and Human Rights* encourages Member States numerous times to support and promote public debate (UNESCO, 2005).

CEST recognizes the importance of involving the public in a democratic process of decision-making, which includes consultations and public debates. According to its mission, moreover, it must inform, raise awareness, gather opinions, foster reflection and organise debates on the ethical issues raised by advancements in S&T. While preparing a position statement, CEST must therefore examine different ways to be informed accurately of the public's opinion on the subject under study.

Characteristics of public consultation

For CEST one of the goals of public consultation is to discover the values, principles and preferences on which public opinion is grounded. Public consultation aims also to verify how a given project responds to the needs and interests of the population or, if necessary, to determine what modifications to make, and what legislative and regulatory frameworks are necessary so that the project obtains the public's support (CEST, 2004b, p. 66).

In its very first position statement, CEST (2004b, pp. 64–65) presented the public consultation as follows:

- There are more than forty different methods (Health Canada, 2000) for integrating citizens in the decision-making process; they must be used according to the subject and the importance of the question, the desired form of the results and the eventual application of the results.
- Public consultations are exceptional places for debates because all actors concerned with the question can participate, thus allowing the emergence of a spectrum of positions and points of view.
- Public consultations, by their nature, ensure the integrity, transparency and honesty of policies which result from them.
- By feeling like actors in the decision-making process, citizens develop a sense of belonging to their community, learn to seek

the common good as opposed to the satisfaction of their personal interests, and gain confidence in government and in the state system as a whole.

On the other hand, when misused, poorly justified or useless (such as when the decision had been made in advance), public consultations contribute to a loss of confidence of citizens in their government and their institutions; for their part, decision-makers see public consultations as a considerable waste of time because citizens do not have, according to them, the capacity or the knowledge necessary to discuss certain subjects with experts – an opinion that CEST does not share, as long as the consultation is supported by an adequate information process. The Commission reiterates the importance of this process, which must provide citizens with adequate information – clear, neutral and complete – on the different aspects of the issue and the various ethical and social problems it raises.

Since this involves 're-establishing the equity of the balance of power between, on the one hand, the promoters, the experts, and the decision-makers who, at the beginning, seem to have all the knowledge and power, and on the other hand, citizens who have less power and knowledge but who are often the people who eventually must support the inconveniences of the projects under discussion' (Beauchamp, 1991, p. 175), the Commission recognizes the pertinence of the following rules for the realization of a public consultation process (CEST, 2004b, p. 68):

- the rules of procedure must be clear and equitable
- the persons leading the consultation must be independent of the political power
- the persons who lead the consultation must be worthy of confidence
- information, as neutral and objective as possible, must be provided before the consultation – an informational period must always precede the period during which the public expresses its opinions
- people who participate in a consultation must be given the means to understand their role and to prepare adequately
- a report on the consultation must be produced following the consultation and before decision-making on the subject, or the publication of a report on the subject

- the time factor is important; a rushed consultation process undermines the process.

Constraints of public consultation

Implementing a process of public consultation is not easy; it requires major human resources and financial investment. Moreover, no method of consultation is perfect. Each has its weaknesses, whether it is the degree to which the population is represented; the heaviness of the mechanism; the high usage of resources, human and budgetary; or the treatment of the results of the consultation and its follow-up. This is true for the most common methods of public consultation, such as public hearings, focus groups and workshops of all kinds, citizen conferences and other panels.

To minimise these difficulties and to consult the population on a scale that is both satisfactory and proportional to its means, CEST experimented with mechanisms, such as online consultations or the organisation of public forums in which experts and citizens-at-large are invited in a context in which the former inform and increase awareness and the latter react by asking questions and offering their commentary and opinions on the subject. In October 2005, CEST held such a forum on the ethical issues of the use of biometry for security purposes, bringing together almost 150 participants from all horizons. Before preparing a position statement, the forum had the goal of initiating public debate on the subject and inviting citizens to make known their views on the matter. In support of the debate, the Commission prepared a consultation document to take stock of the technology and the context of its implementation and to present questions on the ethical issues it raises. In the two cases, the results obtained were not proportional to the efforts dedicated to the realisation of the process and confirmed that it is not easy to convince the population to participate in a consultation exercise, and to express their own views.

It often appears simpler to use opinion polls or surveys to verify the population's level of understanding of a given subject and obtain its reaction to certain consequences that are put forward. For CEST, however, surveys do not respond to the objectives sought by a public consultation. Yet it is a mechanism that can contribute to citizens' level of knowledge and awareness of a given subject. Thus it must be regarded as a stage in the preparation of a public consultation that, after an analysis of the results obtained, makes it possible to better identify what information is necessary, and possibly to target it for the benefit of an actual consultation process.

Nanotechnologies and public consultation

A consultation must be meticulously prepared, and its timing must be carefully chosen. In preparing its position statement on the ethical issues of nanotechnology, CEST decided that a public consultation would be inappropriate before the publication of the position statement. In fact, the most recent surveys addressing questions regarding nanotechnology, whether in Quebec, in Canada or elsewhere, were unanimous: a large majority of the population does not know what nanotechnology is or what it is used for, and has no idea how it could benefit or harm individuals and/or society, to say nothing of its effect on the future of the human species and future generations.

However, considering the importance of consulting citizens on the development of a technology which may have social and economic impacts comparable to those of the industrial revolution or as information technology, the question cannot and should not be overlooked. Also, and without assuming what recommendations the position statement will contain, the Commission could recommend to the government that it hold a large-scale public consultation on the development of nanotechnology in Quebec, and on the ethical issues that may be related to it. The position statement can therefore serve to inform the population about the subject before initiating a consultation in which the Commission could itself participate.

As for its mandate to inform the population, CEST feels that it is important to reach young people, mainly those at the college and university level. For example, following its position statement on genetic databases, the Commission published a brochure on the subject (CEST, 2004a) to make such content more accessible and relevant to the younger generation, particularly concerning genetic data. Moreover, the Commission created a Youth Commission to prepare a position statement on the subject of electronic plagiarism in schoolwork (www.ethique.gouv.qc.ca/eng/activities.html#Information). CEST Youth 2005 is an innovative project created to give young college students (aged 17–20) the opportunity to help in the preparation of a position statement on electronic plagiarism in schoolwork and to formulate recommendations for concerned decision-makers; their work was based on an ethics course in their college study programme concerning ethical issues connected with S&T applications. This project also provided an opportunity for fifteen students among them to be chosen by their peers to see how CEST operates – those chosen met on a weekend to discuss the content of a draft opinion paper accompanied by recommendations which was to be submitted to CEST.

GUIDING POLICY-MAKING

CEST's position statements are part of the global mandate given to it to consult, inform, sensitize and educate the public about ethical issues raised by progress in S&T. They must also take into consideration the second element of its mandate, which is to advise public and institutional leaders about ethical decision-making.

Formulating recommendations

Insofar as possible, CEST's position statements contain recommendations formulated for various stakeholders, be they governmental, institutional or others. These recommendations result from an analysis of the situation – from a scientific, economic and legal perspective – combined with an exploration of the ethical and social issues related to the potential benefits and risks presented by applications of a given technology (for example, nanotechnology).

CEST strives to adhere to a certain number of rules when formulating its recommendations, namely: specify the party to which the recommendation is addressed (at the governmental or institutional level), provide suggestions for the implementation of the recommendation, and specify the outcome sought and the proposed timeline, should this apply.

As already mentioned, CEST sometimes formulates 'cautionary notes' instead of recommendations, particularly when it is impossible to identify the party to whom a recommendation should be directed, to specify a specific objective or to identify ways to achieve a certain outcome. This was the case, for example, in its position statement on GMOs with respect to the instrumentalization of living organisms by using transgenesis: 'In light of its views on the impact of transgenesis on cultural and spiritual representations, the Commission believes that society should develop a means of counterbalancing the current trend toward the commodification of life and averting a certain dehumanization' (CEST, 2003).

The political context is also an element to be considered in the formulation of recommendations, notably regarding the division of powers between different levels of government (particularly federal and provincial), as is the case in Quebec. When the Commission feels that it is important that a recommendation be carried out but that doing so would concern a jurisdiction which lies beyond the Commission's purview and thus in which it cannot directly intervene, the Commission formulates its recommendation such that it demands that the provincial government put pressure on their counterparts at the federal level to carry out that recommendation.

The 'fate' of recommendations

It must be emphasized that an organization such as CEST has no decision-making power, but merely the power to influence. The influence the Commission exerts on decision-makers is greater the more widely its position statements are disseminated and their content is debated with the public.

To intervene at the right time, it is necessary to make provisions for mechanisms to follow the evolution of governmental or institutional policies on the subject of a position statement by making recommendations of an ethical character. It may prove important to recall in parliamentary commissions, notably within a memorandum, certain recommendations concerning important ethical issues and to better inform members of the parliament. Colloquia and other similar activities also can provide platforms for the promotion of recommendations on a given issue.

Finally, it may be useful, if not essential, to contact some of the stakeholders concerned by recommendations formulated in a position statement, so as to convince them of the importance of these recommendations and their relevance for the common good of society.

CONCLUSION

In this chapter, we have tried to show how the early assessment of ethical implications of a technology under development can be challenging. We have also tried to illustrate under what conditions this interdisciplinary endeavour can be achieved successfully in a way that will:

1. provide a framework with a set of principles or values to assess the potential ethical, social, environmental and economic implications of such a technology in a given population

2. provide the information needed to engage in a transparent and informed public debate

3. prevent unjustified fears from mobilizing public opinion

4. provide governments (politicians and policymakers) with some elements that need to be taken into account when formulating their policies.

Finally, it is important to repeat that the dissemination of knowledge and the process of uptake of that knowledge by civil society is an essential component of an ethically - driven decision- and policy-making process.

Each state has to find, based on its culture and traditions, the appropriate way of doing this by involving its own scientists and experts in the discussion and in the preparation of recommendations.

BIBLIOGRAPHY

Beauchamp, A. 1991. L'individu, la collectivité et l'éthique: importance de la consultation publique. *Cahiers de recherche éthique*, Vol. 15, pp. 163–77.

Commission de l'éthique de la science et de la technologie (CEST). 2003. *Pour une gestion éthique des OGM*. Sainte-Foy, Quebec, Commission de l'éthique de la science et de la technologie. http://www.ethique.gouv.qc.ca/fr/ftp/CESTAvisOGMfinal.pdf (Accessed 20 December 2006.)

———. 2004a. *Les Banques d'Information Génétique: « C'est BIG »*. http://www.ethique.gouv.qc.ca/BIG/brochure.pdf (Accessed 20 December 2006.) An English version will be available in 2007.

———. 2004b. *Les enjeux éthiques des banques d'information génétique: pour un encadrement démocratique et responsible.* Sainte-Foy, Quebec, Commission de l'éthique de la science et de la technologie. http://www.ethique.gouv.qc.ca/fr/ftp/ConsultationDocumentFr.pdf (Accessed 20 December 2006.)

Health Canada. 2000. *The Health Canada Policy Toolkit for Public Involvement in Decision Making*. Ottawa, Health Canada. http://www.hc-sc.gc.ca/ahc-asc/alt_formats/cmcd-dcmc/pdf/public-consult/2000decision_e.pdf (Accessed 20 December 2006.)

Khushf, G. 2004. The ethics of nanotechnology: Vision and values for a new generation of science and engineering. National Academy of Engineering (ed.), *Emerging Technologies and Ethical Issues in Engineering*. Washington, DC, National Academies Press, pp. 29–55.

Preston C. J. 2005. The promise and threat of nanotechnology: Can environmental ethics guide us? *HYLE-International Journal for Philosophy and Chemistry*, Vol. 2, No. 1, pp. 19–44. http://www.hyle.org/journal/issues/11-1/preston.htm (Accessed 20 December 2006.)

UNESCO. 2003. *Bioethics: International Implications*. Proceedings of the Round Table of Ministers of Science, Paris, 22–23 October 2001. http://unesdoc.unesco.org/images/0013/001309/130976e.pdf (Accessed 20 December 2006.)

———. 2005. *Universal Declaration on Bioethics and Human Rights*. Paris, UNESCO. http://portal.unesco.org/en/ev.php-url_id=31058&url_do=do_topic&url_section=201.html (accessed 20 december 2006.)

ANNEX 1

LA COMMISSION DE L'ÉTHIQUE DE LA SCIENCE ET DE LA TECHNOLOGIE – CEST QUÉBEC (CANADA)

Description of CEST

The Commission of science and technology ethics (CEST) fosters informed public debate on all ethical issues related to scientific and technological advancement and shares its own thoughts on the issues in the form of position statements, reports or studies submitted to the various stakeholders in the vast field of S&T.

Created in September 2001, CEST is an initiative of the Conseil de la Science et de la Technologie (CST), as provided for in the science and innovation policy *Knowledge to Change the World*. CEST is completely independent, its ties to CST being only administrative.

Guiding principles

The following principles guided the creation of the science and technology ethics Commission (CEST):

- It is a forum for reflection and discussion on the major ethical issues raised by progress in S&T.
- It is a forum for the development and expression of collective choices.
- It gives priority to new problems that have not been settled in other arenas.
- It acts as an information and awareness-raising organisation for the general public, the government, and specialist communities.
- It advises the government and other interested bodies regarding training needs in S&T ethics.

Mission and mandate

Mission

CEST's mission consists, firstly, in informing, raising awareness, gathering opinions, fostering reflection, and organizing debates on the ethical issues raised by advancements in S&T, and, secondly, in proposing general guidelines for stakeholders to refer to in their decision-making.

Mandate

CEST does the following to fulfil its mission:

- Selects its own topics for deliberation or responds to requests by CST or the Minister of Economic Development, Innovation, and Exportation.
- Collects information from national and international sources pertaining to the issues under examination and communicates this information widely through the means it deems appropriate.
- Organizes public debate on these issues, notably through forums, with a view to identifying values specific to Quebec society and developing areas of consensus.
- Takes a stand on these issues, notably in the form of position statements issued to stakeholders in the S&T field, such as practitioners, researchers, political authorities, administrators and members of research or clinical ethics committees.
- Advises the government and other interested bodies on training needs in S&T ethics.
- Annually submits a report on its activities, which is included as an integral part of CEST's annual report to the Minister.
- Establishes formal ties with organizations in the same field abroad.

CEST may also:

- Develop various tools and means to inform the public, the government, and all interested groups about the ethical issues raised by progress in S&T.
- Conduct or commission any studies and research it deems useful or necessary for the fulfilment of its mission.
- Create subcommittees to examine specific issues.
- Develop appropriate mechanisms to consult the general public and specialists on the ethical issues it examines.
- Use all the channels its deems appropriate to communicate the results of its work, including forums, seminars, websites, journals, reports, position statements and studies. While information and consultation activities may be organized in Québec City or Montreal, CEST will pay particular attention to the outlying regions.

CEST membership

Description

CEST is made up of thirteen members and two guest members, divided up as follows:

- one chair, who is also a member of CEST (appointed by the government)
- four scientists (in health, biotechnology and biofoods, the environment, and physical science or new ICT) from the university and industrial communities
- four ethics specialists (philosophers, ethicists, lawyers, social science experts, representatives of the main philosophical or spiritual movements)
- four members from the practicing community (ethics committees, health care administrations, etc.), the media, the education system, or the general public
- two civil servants, sitting as guest members and participating in CEST meetings and deliberations, but without the right to vote.

If needed, CEST can also invite observers who join the Commission in whatever capacity to enrich its work on a particular subject for a given period of time.

Members other than the chair are appointed by the CST for a three-year term, renewable one time only. As for first members, half are appointed for a two-year term, renewable for a maximum of three additional years.

Member selection criteria

Members are appointed after consultation with the scientific and practicing communities. They are selected according to the following key criteria:

- recognition in their community for their credibility and objectivity
- clear interest in and knowledge of S&T ethics
- considerable moral independence from the institution or organization to which they belong

NANOTECHNOLOGIES, ETHICS AND POLITICS – ISBN 978-92-3-104051-1 – © UNESCO 2007

ANNEX 2

SOME EXAMPLES OF APPLICATIONS WITH NANOTECHNOLOGIES

Automotive and aeronautics industries: Nanoparticle-reinforced materials for lighter bodies, nanoparticle-reinforced tires that wear better and are recyclable, external painting and glass that do not need washing, cheap non-flammable plastics, electronics for controls, self-repairing coatings and textiles, catalysts.

Electronics and communications: All-media recording using nanolayers and dots, flat panel displays, wireless technology, sensors, new devices and processes across the entire range of communication and information technologies, factors of thousands to millions improvements in both data storage capacity and processing speeds – and at lower cost and improved power efficiency compared to present electronic circuits.

Chemicals and materials: Catalysts that increase the energy efficiency of chemical plants and improve the combustion efficiency (thus lowering pollution emission) of motor vehicles, super-hard and tough (i.e., not brittle) drill bits and cutting tools, 'smart' magnetic fluids for vacuum seals and lubricants.

Pharmaceuticals, healthcare, and life sciences: New nanostructured drugs, gene and drug delivery systems targeted to specific sites in the body, biocompatible replacements for body parts and fluids, self-diagnostics for use in the home, sensors for labs-on-a-chip, material for bone and tissue regeneration.

Manufacturing: Precision engineering based on new generations of microscopes and measuring techniques, new processes and tools to manipulate matter at the atomic level, nanopowders that are sintered into bulk materials with special properties that may include sensors to detect incipient failures and actuators to repair problems, chemical-mechanical polishing with nanoparticles, self-assembling of structures from molecules, bio-inspired materials and biostructures.

Energy technologies: New types of batteries, artificial photosynthesis for clean energy, quantum well solar cells, safe storage of hydrogen for use as a clean fuel, energy savings from using lighter materials and smaller circuits, nanostructured coatings and coverings, catalysts.

Space exploration: Lightweight space vehicles, economic energy generation and management, ultra-small and capable robotic systems.

Environment: Selective membranes that can filter contaminants or even salt from water, nanostructured traps for removing pollutants from industrial effluents, characterization of the effects of nanostructures in the environment, maintenance of industrial sustainability by significant reductions in materials and energy use, reduced sources of pollution, increased opportunities for recycling.

National security: Detectors and detoxifiers of chemical and biological agents, dramatically more capable electronic circuits, hard nanostructured coatings and materials, camouflage materials, light and self-repairing textiles, blood replacement, miniaturized surveillance systems.

Source: National Science and Technology Council: *Nanotechnology research directions: IWGN Workshop Report*, September 1999, p. xxvii-xxviii.

ANNEX 3

THE MEMBERS OF THE CEST NANOTECHNOLOGY WORKING GROUP

- **Sabin Boily,** physical engineer, nanotechnology consultant, president of the committee, member of the CEST
- **François A. Auger,** Director, Laboratoire d'Organogénèse Expérimentale (LOEX) du CHAUQ, Hôpital du Saint-Sacrement
- **David Carter,** Minister of the Environment of Québec
- **Sylvain Cofsky,** Director of innovation and corporate affairs, NanoQuébec
- **Éric David,** professor, l'École de technologie supérieure, membre of the Chaire de recherche en matériaux et équipements de protection utilisés en santé et sécurité au travail IRSST/ÉTS
- **Édith Deleury,** lawyer, Université Laval, president, CEST
- **André Doré,** public representative having an interest in nanotechnology
- **Denis Godbout,** Office québécois de la langue française, observer and linguistic adviser
- **Peter Grütter,** McGill, Department of Physics; Scientific Director of NanoPic, the innovation platform of nanoscience and nanotechnology of the Natural Sciences and Engineering Research Counsel of Canada (NSERC)
- **Benoît Gagnon,** researcher at the Chaire Raoul-Dandurand en études stratégiques, Université du Québec à Montréal graduate, member of the CEST
- **Mark Hunyadi,** professor of philosophy and applied ethics, Université Laval
- **Michèle S. Jean,** historian, Université de Montréal, president of the IBC of UNESCO, member of the CEST
- **Benoît Lussier,** physicist, Ministère du Développement économique, de l'Innovation et de l'Exportation, observer
- **Teodor Veres,** Canada National Research Council (CNRC), head of the group functional nanometerials

From the Secretariat of the Commission :

- **Emmanuelle Trottier,** bioethics advisor, secretary of the committee (research and drafting)
- **Diane Duquet,** coordinator of CEST (research and drafting, supervision)

The following roundtables and presentations have been organized:

- Presentation on the field of nanotechnology
- Biotechnological aspects of nanotechnology
- Environmental aspects of nanotechnology
- The process of ethical evaluation; the precautionary principle
- Entrepreneurship and security rules and regulations
- Sustainable development and analysis of the life cycle.

ABBREVIATIONS

AFM	Atomic Force Microscope
ALESCO	Arab League Educational, Cultural and Scientific Organization
APCEIU	Asia-Pacific Centre of Education for International Understanding
BSCS	Biological Sciences Curriculum Study
CNT	Carbon nanotubes
CEST	Quebec Commission on Ethics in Science and Technology
COMEST	La Commission mondiale de l'éthique des connaissances scientifiques et des technologies
DOE	Department of Energy
EIBE	European Initiatives for Biotechnology Education
ETC Group	Action Group on Erosion, Technology and Concentration
ELSI	Ethical, legal and social issues
FAO	Food and Agriculture Organization
FDA	Food and Drug Administration
GEC	Genetics Education Center
GEObs	Global Ethics Observatory
GenEd	Genome Education
GMO	Genetically modified organism
HGP	Human Genome Project
HGTNP	Human Genome Teacher Networking Project
HRTEM	High-Resolution Transmission Electron Microscopy
IBC	International Bioethics Committee
ICT	Information and communication technology
ILO	International Labour Organization
IVF	*In vitro* fertilization
IT	Information technology
LCD	Liquid crystal display
LED	Light emitting diode
MDG	United Nations Millennium Development Goal
MEMS	Microelectromechanical systems
MRI	Magnetic Resonance Imaging

NEG	Nanotech Engagement Group
NGO	Non-governmental organization
NHGRI	National Human Genome Research Institute
NIDG	Nanotechnologies Issues Dialogue Group
NIH	National Institutes of Health
NSF	National Science Foundation
OECD	Organization for Economic Co-operation and Development
OST	Office of Science and Technology
PSPD	People's Solidarity for Participatory Democracy
PCR	Polymerase Chain Reaction
R&D	Research and development
RFID	Radio Frequency Identifier
SANi	South African Nanotechnology Initiative
SCCNFP	Scientific Committee on Cosmetic and Non-food Products
S&T	Science and technology
SEM	Scanning Electron Microscope
SPM	Scanning Probe Microscopy
STS	Science Technology Society
STM	Scanning Tunnelling Microscope
SWNT	Single-walled nanotube
TEM	Transmission Electron Microscopy
TWAS	Academy of Sciences for the Developing World
UNDP	United Nations Development Programme
UNESCO	United Nations Educational, Scientific and Cultural Organization
UNHCR	United Nations High Commissioner for Human Rights
WHO	World Health Organization
WIPO	World Intellectual Property Organization
WMA	World Medical Association

Index

Dépôt légal : avril 2007
N° d'impression : 27-0715

Imprimé en France